ILLUSTRATED ENCYCLOPEDIA

OF

SOLID-STATE

CIRCUITS

AND

APPLICATIONS

Donald R. Mackenroth

and

Leo G. Sands

PRENTICE-HALL, INC. ENGLEWOOD CLIFFS, N.J.

(Business and Professional Division)

Prentice-Hall International, Inc., *London*
Prentice-Hall of Australia, Pty. Ltd., *Sydney*
Prentice-Hall of Canada, Ltd., *Toronto*
Prentice-Hall of India Private Ltd., *New Delhi*
Prentice-Hall of Japan, Inc., *Tokyo*
Prentice-Hall of Southeast Asia Pte. Ltd., *Singapore*
Whitehall Books, Ltd., Wellington, *New Zealand*
Editora Prentice-Hall do Brasil Ltda., *Rio de Janeiro*

© 1984, *by*

PRENTICE-HALL, INC.

Englewood Cliffs, N.J.

First Printing

Editor: George E. Parker

Library of Congress Cataloging in Publication Data

Mackenroth, Donald R.
  Illustrated encyclopedia of solid-state circuits and applications.

  Includes index.
  1. Solid state electronics. I. Sands, Leo G.
II. Parker, George E. III. Title. IV. Title: Solid-state
circuits and applications.
TK7874.M2 1984   621.3815′3   83-23077

ISBN 0-13-450537-9

Printed in the United States of America

The Unique, Practical Value
This Book Offers....

Since the invention of the transistor a relatively short time ago, the electronics industry has progressed rapidly in the field of solid-state circuitry. Now, except for a few applications, the role of the electron tube has been entirely taken over by solid-state devices, including discrete (or individual) components such as transistors and diodes as well as integrated circuits which contain hundreds of components on a single so-called "chip." This makes it particularly important for engineers, technicians, experimenters, repair persons and hobbyists to understand basic solid-state circuits, whether these circuits are made up of discrete components, integrated circuits, or a combination of both.

Electronic instruments and systems which at first glance may seem extremely complex, because of the large number of solid-state devices used, are actually quite simple when visualized as "building blocks" of individual interconnected circuits. And that is one of the reasons you will find this book invaluable—it will help you break down overwhelmingly complex electronic systems into basic solid-state circuits that you can then understand more easily.

Although quite often it is not important to know exactly what is contained in an IC, it is important to know the external connections that allow it to operate at peak efficiency, and to know the functions which it is capable of performing. For that reason, this book contains many circuits that utilize ICs, and in many cases the values of the external components in the circuitry surrounding the IC are included also. In addition, many of the circuits utilize only discrete transistors, diodes, and other components, and the explanations accompanying these circuits will help you understand their operation, and even enable you to build your own.

v

The chapters of the book are arranged to progress from simple, basic concepts to more complex ones, so that although each chapter is self-contained for handy reference, the chapters also proceed logically for cover-to-cover or casual reading. And you will find it extremely easy to look up a circuit or an application of a solid-state device. Each chapter contains many circuits which perform similar electronic tasks, so you can compare several RF amplifiers, say, or electronic controls, within a few pages. The circuits in each chapter are listed in alphabetical order and cover an exceptionally wide range of applications. And nearly every entry has at least one accompanying illustration, for handy reference or detailed study.

The first chapter covers semiconductor devices, ranging from the basic diode and junction transistor to such innovations as CMOS and VFET devices. Each type of semiconductor device is described and its characteristics explained in full, practical detail.

The second chapter will also prove extremely valuable, since it explains the characteristics of very basic circuits, such as the use of a transistor in common-emitter, common-base and common-collector configurations. It also details biasing schemes and the basic applications of many solid-state devices. In a sense, this chapter is a bridge between the individual devices found in Chapter 1 and the rest of the book. You can also use it as a reference with added depth to help you further analyze and understand other circuits.

In subsequent chapters you will find: a wide range of audio amplifiers; amplifiers used in both narrow-band and video RF applications; oscillators ranging from the basic kinds of sinusoidal types to those producing outputs that are not sine waves; and more. One chapter gives you a wide range of practical information on AM and FM receivers as well as many of the specialized circuits found in TV receivers. Another entire chapter is devoted to power supplies, presenting everything from simple half- and full-wave rectifiers to the most modern kinds of solid-state switching power supplies. And still another chapter gives you the basic hardware circuitry you need to understand the logic gates found in digital computers, from micros to mainframes.

A chapter on communications devices concentrates on the circuitry used in telephone and other wired communications, while the chapter on measuring and testing devices gives you many practical measuring circuits. There is also a chapter on the radio transmitters used in VHF and CB communications, a chapter on mixers and frequency converters, and a complete chapter dealing with electronic controls. There's even a chapter dealing with practical signal conditioning and interface circuitry, and one chapter each devoted to modulators and to demodulators. At the end of the book you'll find a handy glossary of electronic and semiconductor terms.

You will quickly discover that this is the one book you need to develop a full, clear understanding of the principles and applications of solid-state circuitry. It will not only help you to understand these incredibly versatile devices, but will also enable you to properly use and maintain a wide variety of solid-state equipment. The book is unique because of its organization, heavy use of illustrations, and clear, concise language. It provides the specific data you need, easily and quickly, and you have the assurance this information has been gathered and verified by authors who have many years of experience in designing, using, troubleshooting and maintaining solid-state equipment.

Donald R. Mackenroth

Leo G. Sands

CONTENTS

Introduction • Backward Diode • Diac • Four-Layer
Diode • Hall-Effect Sensor • Insulated-Gate Field-Effect
Transistor • Junction Diode • Junction Field-Effect
Transistor • Junction Transistor • Light-Emitting
Diode • Operational Amplifier • Photodiode • Photon-
Coupled Isolator • PIN Diode • Programmable
Unijunction Transistor • Read Diode • Schottky-Barrier
Diode • Silicon-Controlled Rectifier • Step Recovery
Diode • Triac • Tunnel Diode • Unijunction
Transistor • Varactor Diode • Varistor • VMOS
Power Field-Effect Transistor • Zener Diode

Introduction • Baker Clamp • Bias with Variable
Operating Characteristics • Bootstrapped Emitter
Follower • Cascaded Transistor Amplifier • Common-
Base Transistor Amplifier • Common-Collector Transistor
Amplifier • Common-Drain FET • Common-Emitter
Transistor Amplifier • Common-Gate FET • Common-
Source FET • Current Regulator Diode
Applications • Darlington Circuit • Degenerative
Common-Source FET Amplifier • Emitter Bias • FET
Bias and Parameters • FET Input Stage • Fixed and
Self-Bias • Fixed Bias • Inverting Operational
Amplifier • Noninverting Op Amp • Self-
Bias • Stabilized Bias • Unijunction Transistor as Pulse
Generator • Varactor-Tuned Resonant
Circuits • Varistor Protection • Zener Diode Voltage
Regulation

10. Electronic Controls (*continued*)

Hot Plate Control • Isolated Solid-State Relay • Low-Level Input Sensor • Nixie Tube Driver • Photo Driver for a Triac • PIN Diode Antenna Switch • Solid-State Relay • Thermistor Temperature Control • Time Delay Circuit • Timer with Meter Readout • Touch-Controlled Selector • Touch-Sensitive Switch • Universal Motor Speed Control • Valve or Solenoid Control

Introduction • A-to-D Converter • CMOS Flip-Flop • CMOS Inverter • CSDL NAND Gate • Diode Logic AND Gate • Diode Logic OR Gate • DTL NAND Gate • DTL NOR Gate • D-to-A Converter • ECL OR Gate • Exclusive NOR Gate • Exclusive OR Gate • Flip-Flop • High-Speed TTL NAND Gate • HNIL NAND Gate • Inverter Circuit • LED Driver • LLL NOR Gate • NMOS NAND Gate • NOR Circuit • PMOS NAND Gate • RTL NAND Gate • RTL NOR Gate • Schottky TTL NAND Gate • TTL NAND Gate

Introduction • Automatic Nicad Battery Charger • Cascade Voltage Multiplier • Constant Current Source • Crowbar Circuit • Full-Wave Bridge Rectifier • Full-Wave Rectifier • Full-Wave Voltage Doubler • Half-Wave Rectifier • Half-Wave Voltage Doubler • High-Current Voltage Regulator • Jensen DC-to-DC Converter • Ringing Choke DC-to-DC Converter • Royer Oscillator DC-to-DC Converter • Series Regulator • Series Regulator with Current Limiting • Shunt Voltage Regulator • Step-up DC-to-DC Regulator • Switching Power Supply Principles

Introduction • AM/FM Auto Radio with Mono Audio Output • AM Radio Tuner Using an IC • Automatic

13. Radio and Television Receivers (*continued*)

Frequency Control • Auto Radio/Tape Player • Chroma System • Crystal Set • Discrete Crystal CB Transceiver • Flewelling Receiver • Noise-Operated Squelch • One-FET Radio Receiver • Pressley Impedance Bridge Superheterodyne • Reinartz Receiver • Super-Regenerative VHF Receiver • TV Horizontal Sweep • TV IF Amplifier and Detector • TV Sound Detector • UHF Receiver Preselector • Ultra Audion Receiver • VHF Receiver Front End Utilizing Four MOSFETs • Video Signal Processor • Voltage Doubler Diode Radio Receiver

16. Measuring and Test Instruments (*continued*)

Meter • Microphone Tester • Modulation and Power
Tester • Noise Generator • Radar Detector • Radio
Direction Finder • Signal Tracer • Simple Field-Strength
Meter • Sound Level Meter • SWR Monitoring
Circuit • Tone Squelch Tester

CHAPTER 1

Semiconductor Devices

INTRODUCTION

Strictly speaking, a *semiconductor* is a material that is neither a very good conductor of electricity nor a very good insulator. Conductors such as copper, gold, silver and aluminum have very low resistance, and pass electric current easily. Insulators such as rubber, polystyrene, mica, glass and the like have very high resistance to electric current. Semiconductors, although they can pass an electric current, do not do it well.

1

How, then, has the semiconductor come to be the linchpin of modern electronic circuitry? First, by "doping" a semiconductor material such as germanium or silicon with impurities, we can control its conductive and other electrical properties. And second, by combining one type of doped semiconductor with another, we can create a *junction*, which is the basis for nearly all semiconductor devices as we know them.

Semiconductor materials include germanium, silicon, gallium arsenide, indium arsenide, even copper oxide and lead sulfide. However, the substances most often used in transistors, diodes and other semiconductor electronic devices are germanium and silicon. The individual molecules of these elements exist as crystalline lattices, very structured forms where each molecule is surrounded by and linked to a constant number of other molecules; the linkage is by means of electrons which are held in common by these molecules.

If a crystal of germanium is doped with so-called donor impurities such as arsenic, the bonds between molecules become disarranged, and negatively charged electrons are more free to conduct electric current. At the same time, the arsenic ions, having given up electrons, are positively charged. This type of semiconductor material, with an excess of negative charges (and positively charged donor ions) is called N-type semiconductor. The electrons in N-type semiconductor materials are known as *majority* current carriers, since they carry most of the current.

Now, if a crystal of pure germanium is doped with ions such as those of indium, which act as *acceptors* of negative charges, a similarly altered crystal structure is created. This structure, called P-type germanium, contains negatively charged acceptor ions as well as an excess of what are known as positive *holes*. In P-type semiconductor material, these holes are the majority carriers, and may be thought of as moving throughout the crystal just as electrons move in N-type semiconductor.

If a wafer of germanium (or silicon, or other semiconductor) is doped so that two areas, one of P-type material and one of N-type material, are juxtaposed, a *junction* is formed where the two areas meet. Figure 1-1 shows the two areas and the junction between them.

At the junction, some electrons and holes combine. In addition, holes in the P-type conductor are repelled by the positive charges of the donor ions in the N-type material; and electrons in the N-type material are likewise repelled by the negative charges of the acceptor ions in the P material. A barrier, or depletion region, is thus established at the junction. This barrier, normally on the order of a few tenths of a volt, resists the passage of any particle through the junction.

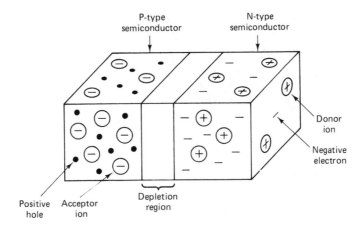

Figure 1-1. Semiconductor Junction

When electricity is applied across the junction, changes occur that form the basis for the operation of the junction diode, the transistor, and other semiconductor devices. Different semiconductor materials and doping agents and levels can produce phenomena that are responsible for the tunnel diode, the varactor, the light-emitting diode, the power rectifier, and other uses of the PN junction. When a semiconductor "sandwich" (PNP or NPN) is constructed, a transistor can be formed. Still other kinds of multiple junctions are responsible for the PIN diode, the four-layer diode, and other useful semiconductor devices.

BACKWARD DIODE

The backward diode, or tunnel rectifier, is often described as a poor tunnel diode. A backward diode can provide some rectification, but its operating characteristic also has an area where forward current does not change or becomes even slightly less even though forward bias is increased. This characteristic is similar to the negative resistance region of the tunnel diode. (See Figure 1-2.)

The schematic symbol of the backward diode is similar to that of a tunnel diode. A backward diode is often placed in series with a tunnel diode so that the normal negative-resistance area of the tunnel diode effectively occurs at a higher voltage.

DIAC

The diac, shown in Figure 1-3, is a three-layer, two-junction device that essentially acts as a pair of back-to-back diodes. No matter what polarity of the voltage is applied to the diac, one of the two PN junctions will be reverse-

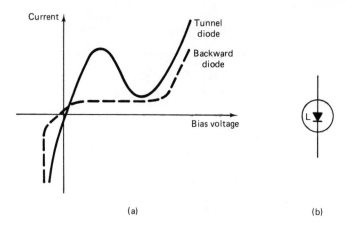

(a)   (b)

Figure 1-2. Backward Diode

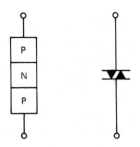

Figure 1-3. Diac

biased, and only a small leakage current will flow. As the voltage is increased, it will ultimately reach the *breakover* point; here, the reverse-biased junction avalanches, and the device exhibits a negative-resistance characteristic similar to that of a tunnel diode.

The unique feature of the diac is that its negative-resistance characteristic is symmetrical—that is, it occurs regardless of the polarity of the applied voltage. Diacs are used in triggering and current reversal circuits.

FOUR-LAYER DIODE

The four-layer, or Schockley, diode is created from a pair of PN junctions joined together, as shown in Figure 1-4(a). Voltage is applied so that the center NP junction is reverse-biased, while the outer PN junctions are forward-biased. This results in the characteristic curve shown in Figure 1-4(b).

As the voltage is increased, the device conducts hardly at all until a threshold (the *breakover* voltage) is reached. Then, as the center NP junction avalanches, it conducts readily and its resistance falls. The four-layer diode

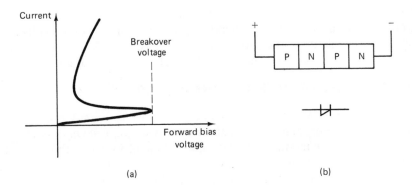

(a)   (b)

Figure 1-4. Four-Layer Diode

is used most often in switching circuits. It can generally switch faster and handle higher voltages than can the unijunction transistor.

HALL-EFFECT SENSOR

In 1879, E.H. Hall discovered that when a conductor carrying a current is placed in a magnetic field, a voltage appears across the conductor that is perpendicular to both the conductor and the lines of flux of the magnetic field. With normal conductors these so-called Hall voltages are very small, but semiconductors produce voltages that are usable in instrumentation and control circuits, and for detecting and measuring magnetic fields.

Figure 1-5 shows a typical Hall-effect sensor. A very thin wafer of germanium, indium arsenide, or other semiconductor is attached to a source of control current and placed in a magnetic field. A Hall voltage, typically

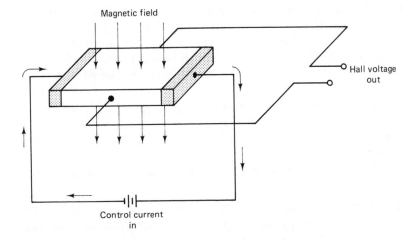

Figure 1-5. Hall-Effect Sensor

on the order of 0.5 millivolts or less, is produced at right angles to both the magnetic field and the control current path. The magnitude of voltage developed is dependent on the intensity of the magnetic field, the amount of control current, and the thickness and type of the semiconductor wafer.

INSULATED-GATE FIELD-EFFECT TRANSISTOR

The insulated-gate field-effect transistor (IGFET) is similar to a junction field-effect transistor; and both are similar to a vacuum tube in that a voltage on a *gate* controls current flow through a channel in much the same way grid bias controls current flow between cathode and plate. However, in the IGFET, the gate is insulated from the rest of the device.

Most of the insulated-gate FETs in use are constructed using metal-oxide silicon (MOS) technology. Although discrete FETs are in wide use, three processes, PMOS, NMOS and CMOS, lend themselves to use in large-scale integrated circuitry as well.

PMOS Circuitry: Figure 1-6(a) shows a cross section of a typical PMOS field-effect transistor, including the source (S), the insulated gate (G), the drain (D) and the substrate or body (B).

The body of the transistor is N-type silicon in which two P-type silicon regions have been inserted. The gate is actually an electrode that is positioned between these two P layers and insulated from the body by a thin layer of silicon dioxide.

A voltage applied to the gate changes the conducting properties of the surface of the N-type substrate beneath it. As the gate voltage is raised past a critical point, called the threshold, a P-type enhancement region, or *channel*, for current to flow is created between the source and drain regions.

In the off state, the gate voltage is kept low or below the threshold of the FET. When a negative potential is applied to the gate, however, the surface begins accumulating mobile positive charges that are attracted by the negative gate. If the gate voltage is sufficiently negative, the substrate surface is essentially changed to P-type silicon with a continuous layer of mobile positive charges forming a channel between drain and source. Now current can flow freely through this newly created channel, and the FET is turned on.

When the channel is formed by the application of the gate voltage, the device is an *enhancement mode* FET. When the channel is created by the difference in potential between drain and source, the FET is a *depletion mode* device.

PMOS circuitry is in wide use because it is simple, reliable, and comparatively easy to manufacture.

NMOS Circuitry: The NMOS FET is made up of two regions of N-type silicon inserted into a P-type substrate. The gate (G) is an electrode positioned between these two N-regions and insulated from the body by a thin layer of silicon dioxide, as shown in Figure 1-6(b).

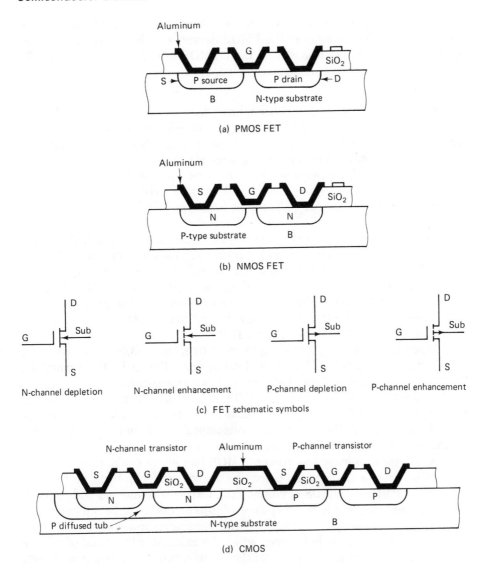

Figure 1-6. Insulated Gate Field-Effect Transistor

When the gate voltage is low, no current flows between the source (S) and the drain (D), and the transistor is turned off. However, when the positive potential on the gate is increased past a threshold, negative charges (electrons) accumulate on the surface of the substrate, and that area below the gate essentially becomes N-type silicon. This newly created channel permits current to flow between source and drain, and the FET is turned on. Like PMOS FETs, NMOS channels can be either enhancement mode or depletion mode. The schematic symbols for these devices are shown in Figure 1-6(c).

The mobility of the electrons in the NMOS FET is about three times that of the positive charges in the PMOS device, so NMOS circuitry can be smaller, faster, or use less power than PMOS circuits. However, NMOS circuitry is more difficult and expensive to manufacture than that constructed using PMOS.

CMOS Circuitry: In CMOS (complementary metal-oxide semiconductor) circuitry, the basic "building block" is a pair of adjacent field-effect transistors—one P-channel and one N-channel—on a single N-type substrate. The transistors are produced by inserting a large P-type region into the substrate in addition to the P source and drain regions for the P-channel FET. As shown on the left in Figure 1-6(d), this large P-type region acts as the substrate for the N-channel FET; the N source and drain regions are added later.

CMOS requires very little power—in a typical electronic calculator, for example, only about 5 μW of standby power are used as opposed to 400mW or so of battery power consumed when the calculator is operating.

In operation, when the gate voltage is positive, the CMOS circuit is turned on. The NMOS transistor is turned on because the gate voltage is past threshold, but the PMOS FET remains off because the gate voltage for it is below threshold; thus no current flows.

When the common gate voltage is negative, the CMOS circuit is turned off; that is, the NMOS FET is turned off and the PMOS FET is turned on. Once again, no current flows between the two transistors.

Separately, a PMOS and an NMOS FET would each consume power. But together in CMOS, only one transistor is on at a time, and there is virtually no current flow or power consumption. The only power is used during the switching of the circuit from on to off.

Although requiring less power, CMOS has lower packing density (the number of circuits that can be placed on a given size chip) and higher cost than either PMOS or NMOS circuitry. Still, CMOS is found in a wide variety of applications, especially where power requirements must be minimized. In a hand-held calculator, for example, CMOS circuitry allows program or data storage memory to remain on even when the calculator is turned off. Since the power drain measures only tenths of microamps, program and data storage are virtually permanent.

JUNCTION DIODE

The junction diode is one of the simplest types of semiconductor devices. A junction diode is formed by the junction of P-type silicon or germanium and N-type silicon or germanium, as shown in Figure 1-7. This junction is created when the P-type material and the N-type material are joined in a single crystal.

The critical area for the operation of a junction diode is in the region of the junction itself. If reverse bias is applied to the diode, as shown in

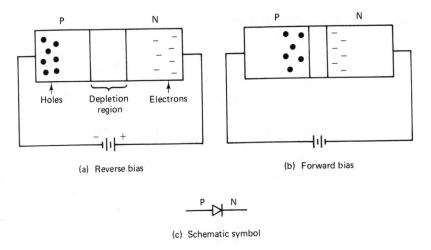

(a) Reverse bias   (b) Forward bias

(c) Schematic symbol

Figure 1-7. Junction Diode

Figure 1-7(a), the majority carriers (the holes) in the P-type material are attracted toward the negative battery terminal, while in the N-material the electrons are repelled from the junction and attracted to the positive terminal. This effectively makes the depletion region at the junction wider, and the diode is even more resistant to current flow in this direction.

In the case of forward bias, as shown in Figure 1-7(b), the majority carriers (holes in the P-material and electrons in the N-material) are repelled from the battery terminals into the area of the junction. There they recombine heavily, effectively reducing the width of the depletion region barrier and readily permitting current flow through the device.

As with any diode, then, a junction diode permits current flow in one direction, but resists current flow in the opposite direction. If the reverse bias is high enough, of course, current can be made to flow even in the reverse direction; this potential is called the *breakdown, avalanche,* or *Zener* voltage, and exceeding it can, depending on the type of diode, cause permanent damage to a conventional junction diode. (Zener diodes are normally operated in this range.)

Junction diodes are often rated according to the following electrical characteristics:

Reverse Voltage or *Peak Inverse Voltage*: Maximum reverse bias that can be applied to the diode before permanent damage occurs.

Forward Voltage: Maximum forward bias voltage at a given current.

Reverse Current: Also called Back Current. When reverse bias is applied, this is the maximum average reverse current that can be sustained by the diode without damage.

Reverse Recovery Time: The time required for a diode to recover its reverse blocking action after a surge of reverse current.

JUNCTION FIELD-EFFECT TRANSISTOR

The junction field-effect transistor (JFET) is made up of a channel of one type of semiconductor material diffused into a block of another type. Current through the channel is controlled by voltage applied to a gate, in much the same way grid bias in a vacuum tube controls current flow between cathode and plate.

Figure 1-8 illustrates the construction of an N-channel JFET. Two electrodes of N-type semiconductor (usually silicon) and a channel between them are diffused into a substrate of P-type semiconductor. A *gate* of P-type semiconductor is added over the channel between the source and the drain. (In practice, the gate is usually connected electrically to the substrate.) When a source of electromotive force is applied to source and drain, current (I_{DS}) flows through the channel because of the majority carriers, the electrons, in the N-type silicon. Current flow is from source to drain. As voltage is increased, current through the channel increases until the region is saturated and no further increase in current flow is possible.

If the gate-source PN junction is then reverse-biased (that is, a negative potential applied to the P-type gate), this bias causes a depletion region to form in the area of the junction, effectively reducing the size of the channel for I_{DS} current flow. The gate, then, acts much like the grid in a vacuum tube. The cutoff (or sometimes, the pinch-off) voltage is the value for this gate voltage that cuts off entirely the flow of current through the channel.

Unlike the transistor, the FET is controlled by the gate voltage—little gate current flows. For this reason, FETs have very high input impedance.

The JFET shown is an N-channel device; in a P-channel FET, the areas of N- and P-type semiconductor are reversed, and conduction through the P-channel occurs because of majority-carrier holes. Biasing, of course, is opposite to that of the N-channel FET, but otherwise operation is virtually the same. N-channels have better conductivity than P-channels, so N-channel FETs are usually faster and more efficient.

JUNCTION TRANSISTOR

A junction transistor is a semiconductor formed with a layer of N material sandwiched between two P layers, as in the PNP transistor shown in Figure 1-9(a), or of a P layer placed between two N layers, as in the NPN transistor shown in Figure 1-9(b).

We will consider first the action of the PNP transistor. In the PNP transistor, one of the P layers forms the *emitter*; this is always shown by the arrow on the schematic symbol for the device. (The symbol for a PNP transistor has the arrow Pointing iNto the Plate, while in the symbol for the NPN transistor the arrow is Not Pointing iN.) The central N layer is much thinner than the one two regions; it forms the *base* of the PNP transistor. The third layer, made up of P material, is the *collector*. Thus the three layers form two PN junctions.

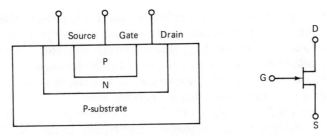

Figure 1-8. Junction Field-Effect Transistor

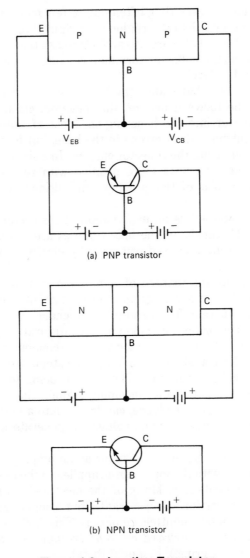

(a) PNP transistor

(b) NPN transistor

Figure 1-9. Junction Transistor

In normal operation, the base-collector junction is reverse-biased. With the emitter circuit open and no bias applied to the emitter junction, a certain collector-base leakage current, I_{CBO}, flows between collector and base. (This may also be called the collector cutoff current.) The leakage current occurs primarily because of the minority carriers in the two regions, and is relatively constant over a wide range of collector reverse bias.

When forward bias is applied to the base-emitter junction, emitter current (I_E) flows readily through the base-emitter region. In the PNP transistor, this current is due primarily to the flow of holes from the P-type emitter into the N-type base.

The base region is very thin, often only a few molecules wide, so most of the holes flowing into the base region pass right through it into the base-collector junction. These holes are attracted through the collector toward the negative potential at the collector terminal, causing an increase in collector current above the I_{CBO} level.

A signal condition that makes the base more negative with respect to the emitter aids the forward bias on this junction, causing more holes to flow from the emitter into the base region. Most of these holes from the emitter do not combine with electrons in the base, but instead flow into the collector region, increasing the collector current. In a similar fashion, a signal condition that makes the base more positive with respect to the emitter opposes the forward bias of this junction, and ultimately reduces collector current.

This, much simplified, is an explanation of transistor action in a PNP device. For an NPN transistor, the bias voltages are of opposite polarities, and the flow of current is caused primarily by electrons diffusing into the base-collector junction.

The most significant feature of a junction transistor is that it is a *current-operated* device: a change in emitter current can cause a similar change in collector current. The ratio of these changes is the current gain, or *alpha*, of a transistor (in a common-base configuration), and is given by the formula $\alpha = I_C/I_B$, where collector voltage remains constant.

The value of the transistor as an amplifier stems from the impedances of the emitter and collector circuits. The emitter, being forward-biased, presents a low-impedance path to an incoming signal. The reverse-biased collector, because of its high impedance, can be fed into a high-resistance load. This allows a small change in input voltage to generate a large change in output voltage.

The most common use of a transistor as an amplifier is in a common-emitter configuration, with the input signal applied to the base and the output signal picked off the collector. Many of the electrical parameters listed in catalogs and data sheets for specific transistors assume that the transistor is connected as a common-emitter amplifier. This includes the *gain* of a transistor, also called the *forward current transfer ratio*, which may be symbolized by α' (alpha prime), h_{FE}, or β (beta).

A transistor may also be used as a switch, in which case its operation is swung quite rapidly from the *cutoff* state (little or no collector current) to the on state. Transistors are sometimes operated in *saturation*, too. In saturation, both junctions (including the base-collector junction) are forward-biased, and the transistor conducts so heavily that a change in emitter current has no effect on a change in collector current. Transistor cutoff and saturation are the basis for most digital circuits.

LIGHT-EMITTING DIODE

The light-emitting diode, or LED, is a semiconductor that emits enough visible light to be used in calculator displays or as an indicator light. The principles behind the LED were observed as early as 1907, when an experimenter, H.J. Round, found that when wires from a battery were touched to a crystal of silicon carbide, yellow light was produced at the junction of one of the wires. In fact, with the proper voltage, any ordinary semiconductor diode gives off a small amount of infrared light. However, most commercial LEDs are constructed of gallium arsenide (GaAs) or gallium arsenide phosphate (GaAsP) because these substances are relatively efficient and give off light that is visible to the human eye.

The LED, shown in Figure 1-10, operates much as any normal diode. It consists of an N-region of semiconductor material joined to a P-region. When enough forward bias is applied to the junction, electrons in the N-region gain enough energy to migrate across the junction to the P-region, and in the LED this energy is given off as light as the electrons combine with holes. This light may also be called *recombination radiation*.

During manufacture the LED is ground to a domed surface or covered with a lens. This makes the device more efficient, allowing most of the light to be emitted instead of being reflected back into the junction.

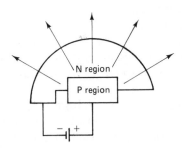

Figure 1-10. Light-Emitting Diode

OPERATIONAL AMPLIFIER

The original operational amplifiers were hot, bulky, high-voltage tube circuits that were used to perform mathematical computations in early computers. A perfectly serviceable op amp could still be constructed using vacuum tubes, of course, but it is semiconductors and integrated circuits that have made the small, inexpensive operational amplifier ubiquitous in all realms of analog signal-handling.

What is usually called an operational amplifier is, strictly speaking, a *differential amplifier*. As shown in Figure 1-11(a), the device has two inputs and a single output. One of the inputs, the (+) or noninverting, is in phase with the output, while the other input, the (−) or inverting, is 180° out of phase with the signal at the output. The two inputs are shown at the base of a triangle that represents the entire amplifier, and the output is shown taken off the opposite point. Additional connections for compensation and offset are sometimes shown along the sides of the triangle.

For the most part, an op amp can be considered merely as a "black box," without consideration for its internal circuitry. However, Figure 1-11(b) illustrates the internal functions of a basic transistorized operational amplifier.

Both the inverting and noninverting inputs are fed to a differential stage which consists of parallel common-emitter amplifiers Q1 and Q2. Transistor Q3 provides constant-current source biasing for Q1 and Q2. Since current through R5 and the collector of Q3 is held constant, any input to, say, Q1, that increases the current flow through that transistor must result in reduced current flow through Q2. Q1 and Q2 thus form a *difference amplifier*, which amplifies the difference between the signals applied to the inverting and noninverting inputs.

The outputs from the collectors of Q1 and Q2 are fed directly to the bases of Q5 and Q4, respectively, for more amplification of each signal. The output of this second stage is single-ended, being taken only off the collector of Q5. After further amplification by Q6, the signal is applied to the output stage, consisting of transistors Q7 and Q8 connected in a push-pull, complementary-symmetry configuration.

The change in output when one or the other input is varied is known as *difference* gain, or differential gain. With no feedback applied, this is also known as *open-loop* gain. Ideally, all gain in an operational amplifier would reflect only the difference between the two inputs, but this is seldom the case.

In an ideal op amp, the output voltage would depend only on the difference between the inputs, and not on the actual values for their voltages. Thus, an op amp with both inverting and noninverting inputs of 1.5 volts should give exactly the same output as with inputs of 15 volts.

Actual op amps, however, have a certain amount of common-mode gain. Common-mode gain is a change in output when the inputs, although equal,

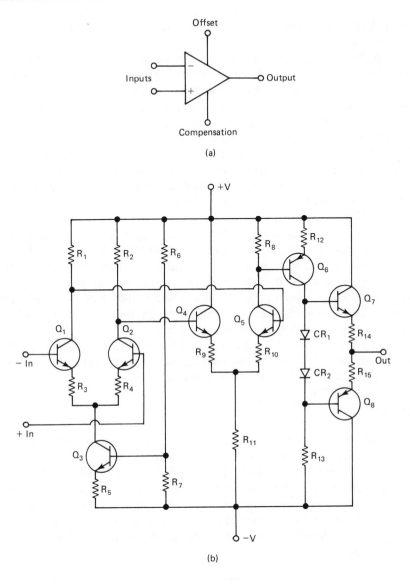

Figure 1-11. Operational Amplifier

are varied over a range. The difference between the desirable difference gain and the undesirable common-mode gain is the measure of the *common-mode rejection ratio* for an operational amplifier:

$$CMRR = \frac{\text{Difference gain}}{\text{Common-mode gain}}$$

In general, the higher the figure for CMRR, the better the amplifier. Typical CMRR values are 70-100 dB.

With the two inputs shorted together, an ideal op amp would have an output of 0. Because of differences in balancing among internal components, real-life op amps have an output under these conditions. This output is called the *offset*. It can be corrected by applying an offset voltage or current at the input, or at separate pins provided on some op amp ICs. The offset value also changes with temperature, and this is called the *drift* of the op amp.

Another correction that must often be applied to an operational amplifier is *compensation*. Compensation can be in part of the internal circuitry of the device, or it can be applied externally in the form of a capacitor or RC network. Compensation helps correct drift, but its primary purpose is to prevent oscillation of the amplifier.

Because of internal capacitance and other figures, when a sharply varying signal is applied across the inputs, the output of an op amp does not track it perfectly. The *slew rate* of an operational amplifier is a measure of this tracking; it is defined as the rate of change in output voltage with a sharply changing, large-amplitude input voltage. Slew rate is usually given in V/μsec. The higher the slew rate for an operational amplifier, the better its high-frequency or high-amplitude operation.

The ideal operational amplifier would have infinite bandwidth, infinite input impedance, zero output impedance, and infinite voltage gain. Real-world op amps, of course, are another matter.

PHOTODIODE

The photodiode is a semiconductor that responds to light energy. A photodiode, whose schematic symbol is shown in Figure 1-12, is normally operated with reverse bias. With no light applied, a very small reverse current flows through the PN junction because of the minority carriers present. When light strikes the photodiode, however, it causes a great increase in the concentration of minority carriers, and reverse current through the photodiode rises sharply. The increase in current is almost directly related to the intensity of light.

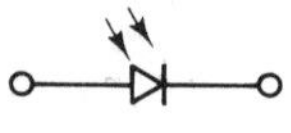

Figure 1-12. Photodiode

PHOTON-COUPLED ISOLATOR

A photon-coupled isolator, like the General Electric 4N40 shown in Figure 1-13, is actually a gallium arsenide LED and a light-activated transistor, SCR, or other semiconductor packaged together. In the case of the

Figure 1-13. Photon-Coupled Isolator

4N40, current through the diode throws light on the SCR and triggers it just as a gate pulse triggers a normal SCR. The photon-coupled isolator is often used in computer interface and control circuitry, where a high degree of isolation of power-carrying circuitry from sensitive logic or signal circuits is desired. The photon-coupled isolator is also known as an optical coupler or optoisolator. (*Courtesy of General Electric Company.*)

PIN DIODE

The PIN diode contains a thin layer of *intrinsic* (I) high-resistance conducting material placed between the P and N materials (see Figure 1-14). When reverse-biased at microwave frequencies, the device is effectively a high-resistance capacitive network. When forward bias is applied, the I-region becomes a conductor because of the holes and electrons it absorbs from the neighboring P and N regions. The forward resistance of the device actually decreases with increasing current, much like the resistance of a tunnel diode. The PIN diode can generally handle more power than a tunnel diode, and is often found operating at microwave frequencies as a modulator or DC-controlled switch.

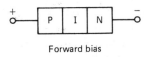

Forward bias

Figure 1-14. PIN Diode

PROGRAMMABLE UNIJUNCTION TRANSISTOR

The programmable unijunction transistor (PUT) is a four-layer device that is similar to a silicon-controlled rectifier. In the PUT, however, the anode gate is available for triggering. (Compare the schematic symbols for the two devices in Figure 1-15.) The triggering voltage of the PUT is programmable, and can be selected by a voltage divider network.

The PUT is faster and more sensitive than a normal unijunction transistor. It is often used in phase control and timing circuits.

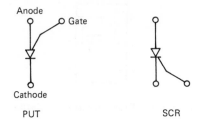

Figure 1-15. Programmable Unijunction Transistor

READ DIODE

The Read diode was developed at Bell Telephone Laboratories by W.T. Read. It is a PNIN device, where the third layer (I-layer) is a band of *intrinsic*, normal conducting material. As bias on the device is raised from a state of reverse leakage, a combined tunneling and avalanche effect occurs. Like the negative-resistance characteristic of a tunnel diode, this tunneling can be used in an oscillator or switch. However, the power levels in the 10 GHz to 50 GHz range are much higher in the Read diode than in a normal tunnel diode.

SCHOTTKY-BARRIER DIODE

When a thin film of certain metals, such as gold, is deposited on a semiconductor surface, a Schottky barrier is formed. This barrier, or junction, forms a diode in a manner similar to a PN junction; however, the electrical characteristics of the junction are such that Schottky-barrier diodes can switch very rapidly from forward conduction to reverse blocking. Current through an SBD also rises much more rapidly as forward bias is increased than does current through a normal PN junction diode. The Schottky-barrier diode is also known as a "hot-carrier" diode.

SILICON-CONTROLLED RECTIFIER

A silicon-controlled rectifier, as shown in Figure 1-16, appears to the circuit as a rectifier diode with a gate; even though forward bias is applied to the diode, no current flows until a positive potential is also felt at the gate.

The SCR is a type of thyristor. It is constructed with four layers (PNPN) of semiconductor material, resulting in three junctions. Diode junctions P1N1 and P2N2 are forward-biased when a positive potential is applied to the anode. However, with no potential on the gate lead, junction N1P2 is reverse-biased, and the device remains cut off.

When the proper positive gate voltage is applied to P2, it causes a small base current to flow in the transistor formed by N1, P2 and N2. This in turn

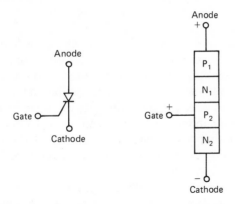

Figure 1-16. Silicon-Controlled Rectifier

causes a large emitter-collector current to flow as in an NPN transistor, with N2P2 acting as the emitter-base junction and N1P2 forming the collector-base junction. In addition, the transistor formed by P1, N1, and P2 also develops heavy collector current, with the result that the entire device is turned on, and heavy current flows from cathode to anode. Even though gate voltage is now removed, the device stays on until the anode voltage is lowered or reversed in polarity.

The SCR switches relatively slowly, so its use is limited to low-frequency applications. However, it can handle large amounts of both voltage and current, and is often found in motor controllers, power supplies, and other high-power applications.

Among other parameters, SCRs are rated according to:

Peak Reverse Blocking Voltage: Maximum voltage that can be applied between anode and cathode with the device turned off.

Gate Trigger Voltage: The minimum DC voltage between gate and cathode required to produce gate current and turn the device from off to on.

Forward Current: Maximum forward anode current that the SCR can handle.

Holding Current: The minimum anode current required to hold the SCR on after gate voltage is removed.

STEP RECOVERY DIODE

A step recovery diode is similar to a varactor in that it is a device formed with a graded junction. However, the capacitance of the step recovery diode is designed to return quickly to a selected value when the diode is switched. This characteristic gives the SRD excellent harmonic generation capabilities, and the step recovery diode is often used in multipliers, especially in the microwave range. Its schematic symbol is usually the same as for a varactor.

TRIAC

The triac is essentially two SCRs, connected back to back and using the same gate. The device can conduct in either direction, and is constructed so that a gate potential of either polarity can turn the triac on.

Figure 1-17 illustrates the schematic symbol for a triac. As with an SCR, once the triac has been gated on, the gate has no more control. However, because the triac can conduct in both directions, it cannot be turned off as an SCR often is, by simply reversing the polarity of the voltage applied to the anode. In a triac, current flow must fall to around zero for some minimum time in order for the device to remain off.

The triac can be turned on with a positive or a negative (with respect to anode 2) gate pulse, and the potential at anode 1 can likewise be positive or negative. The triac is most often used for current switching in AC circuits.

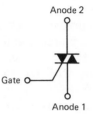

Figure 1-17. Triac

TUNNEL DIODE

The principles of the tunnel diode were revealed by a Japanese physicist, Leo Esaki, in 1957, and the device is sometimes called the Esaki diode. Its most significant feature is that it has an area of negative resistance, where an increase in forward bias produces a decrease in current flow.

In a normal PN rectifying diode, the semiconductor materials are lightly doped with impurity atoms. If these semiconductor materials are more heavily doped, the depletion region (the junction) between the P and the N material is much thinner, yielding a diode with electrical characteristics that are markedly different from those of an ordinary diode. This ultra-thin depletion region is found in the tunnel diode.

As shown in Figure 1-18(a), a normal diode can be reverse-biased over a wide range before current flow occurs; but the junction region of a tunnel diode is so thin that breakdown occurs almost immediately upon application of reverse bias.

When *forward* bias is applied to a tunnel diode, however, and increased from zero volts upward, the diode is extremely conductive until a level of *peak current* is reached. This high current with relatively low forward bias is caused by a phenomenon known as electronic *tunneling* through the ultra-thin barrier of the diode.

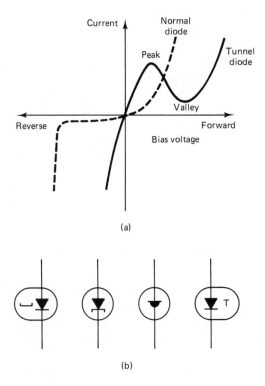

(a)

(b)

Figure 1-18. Tunnel Diode

As forward bias is increased past the peak, progressively less tunneling occurs, and current actually drops until a *valley current* level is reached. If forward bias is increased still more, current again rises, this time in the same manner as in a normal PN junction diode. This current is called *injection current.*

The levels of forward bias voltage producing peak current are very low, typically 50-250 millivolts. As switches, tunnel diodes are very fast, and are often used in high-frequency and microwave circuitry. Figure 1-18(b) shows some typical schematic symbols used for the tunnel diode.

UNIJUNCTION TRANSISTOR

A unijunction transistor is a two-layer device with three terminals, as shown in Figure 1-19. It is sometimes called a double-base diode, since two of its three terminals are connected to a base region. The third terminal is connected to the emitter.

In practice, the UJT is usually wired with one base to ground or common and the other base to a source of forward bias voltage. With the emitter disconnected, the base acts as a voltage divider and permits a portion of the source voltage to appear at the emitter.

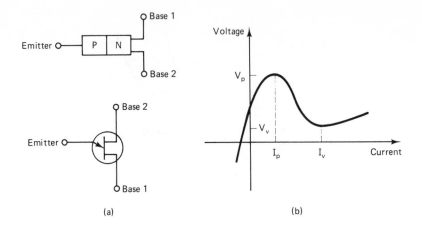

Figure 1-19. Unijunction Transistor

If the UJT is wired so positive emitter bias is also applied, the emitter will be reverse-biased as long as this bias is less than the portion of source voltage that is already present at the emitter. However as emitter bias is made more positive, the emitter becomes forward-biased and emitter current flows.

Like the tunnel diode, the unijunction transistor exhibits a negative resistance characteristic: that is, at a certain peak level of emitter current (I_p), as emitter current increases, bias voltage decreases to a valley point before increasing again.

The UJT is a fairly low-frequency device that is often used in multivibrators, sweep and pulse generators, comparators, and switching circuits.

VARACTOR DIODE

A voltage variable capacitance diode, or varactor diode, is actually a semiconductor device whose capacitance varies depending on the voltage applied. A varactor can thus be substituted for a variable capacitor in many cases, providing solid-state tuning of circuits.

During construction, the depletion region of a PN semiconductor junction is doped so as to provide a graded junction—that is, a junction whose concentration of doped impurities is varied but controlled throughout to tailor the electrical characteristics of the device.

Figure 1-20 shows typical schematic symbols for varactors.

VARISTOR

A varistor is a device, usually made of metallic oxide, that acts as a pair of back-to-back Zener diodes. With no or low voltage applied, the varistor exhibits a high characteristic impedance. In the case of a high-voltage tran-

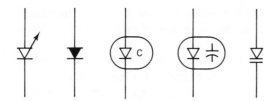

Figure 1-20. Varactor

sient, however, the varistor presents a low-impedance path and clamps the input voltage to a safe level. Varistors are used in protecting voltage-sensitive components and circuitry from transient pulses. Figure 1-21 shows the schematic symbol for a varistor.

Figure 1-21. Varistor

VMOS POWER FIELD-EFFECT TRANSISTOR

The VMOS (vertical-groove metal-oxide semiconductor) FET is similar to a conventional MOSFET in that a gate controls current flow through the channel between source and drain. As shown in Figure 1-22 however, the VMOS FET is constructed with a V-shaped channel. Current flow in the channel is thus *vertical* (instead of along a horizontal surface as in the conventional MOSFET), resulting in much greater current-handling capabilities. The VMOS FET is used in high-current, high-voltage applications where the high input impedance and fast switching time of a field-effect transistor are required.

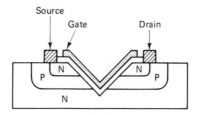

Figure 1-22. VMOS Power Field-Effect Transistor

ZENER DIODE

If an increasing reverse bias is applied to a junction diode, little reverse current flows until the *Zener,* or *avalanche,* voltage is reached. However, at the Zener point the reverse current suddenly increases as the junction breaks down. This sudden increase of current can be seen in the graph in Figure 1-23.

Although breakdown can be destructive to some diodes, the Zener diode is designed specifically to operate in this range. When a reverse-bias voltage is placed across a Zener diode, negative current increases sharply as the Zener voltage is reached. This increased current across an external load increases the voltage drop across the diode junction; the net effect is that the Zener diode provides constant voltage (and constant current) across the load.

Zener diodes are most often used as voltage reference devices, to "clamp" voltages to specified levels. They are also found in overload and surge protection circuitry, and in clippers and other shaping circuits.

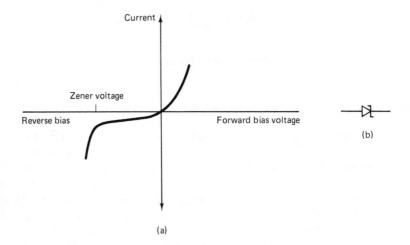

Figure 1-23. Zener Diode

CHAPTER 2

Basic Circuit Configurations

INTRODUCTION

Although electronic components, including semiconductors, are sometimes used in seemingly mysterious and arcane ways, for the most part they are interconnected in patterns of simple circuits. The most complex radar system, the most powerful computer, uses the same principles that are found in a crystal radio set or a basic transistor amplifier. Whether you are designing, troubleshooting, or modifying an electronic device, you will find the task much easier if you visualize it as simply a group (or hundreds, or thousands) of interconnected more basic circuits.

25

Transistors, for example, may be connected in one of three basic configurations: common-emitter, common-base, or common-collector. Field-effect transistors are found in similar patterns: common-gate, common-source, or common-drain.

All semiconductor devices are sensitive to heat, and their performance is severely affected by the heat that must be dissipated in a circuit and also by the ambient temperature of the surroundings. In a transistor or FET amplifier, this means that not only must techniques be used that properly bias the device to operate on the desired portion of its characteristic curve, but biasing or other circuitry must also act to reduce the temperature sensitivity of the semiconductor.

Another consideration affecting the biasing of a semiconductor device is whether it is used as a simple amplifier or as a switch. For example, when used as an amplifier, a transistor is biased so its emitter-base junction is forward-biased and its collector-base junction is reverse-biased. As a switch, a device is usually biased near cutoff (when no or minimal current flows through the semiconductor junction), or even near saturation; when a transistor goes into saturation, both PN junctions are forward-biased, and current through one does not affect current flow through the other.

As for operational amplifiers, the signal input may be connected to the inverting ($-$) or noninverting ($+$) input pin. Similarly, the feedback loop can be hooked from the output to either input, depending on whether the op amp is being used as an amplifier or as an oscillator.

This chapter shows many basic applications and biasing for semiconductor devices.

BAKER CLAMP

The switching time of a transistor can often be improved if the transistor is prevented from going into saturation. In saturation, the input signal drives the transistor so that the base-collector junction becomes forward-biased.

One way to prevent saturation is by adding Baker clamp diodes, as illustrated by D1, D2, and D3 in Figure 2-1. The forward drop across a diode is about 0.6 volt. D1 and D2 connected in series ensure that the base of transistor Q1 is always 1.2 volts below the input, while D3 ensures that the collector is only one diode drop below the input. This means that the collector of Q1 will always be at least 0.6 volt positive (hence, reverse-biased) with respect to the base, and the transistor will never go into saturation.

A Darlington configuration can achieve the same result: even though the drive transistor goes into saturation, the output transistor does not.

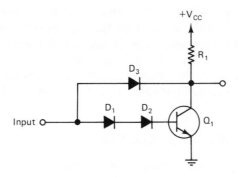

Figure 2-1. Baker Clamp

BIAS WITH VARIABLE OPERATING CHARACTERISTICS

Figure 2-2 shows a common-emitter amplifier that uses a voltage divider and an emitter resistor to obtain both fixed bias and self-bias for the transistor. The voltage applied to the base of Q1 is reduced by the divider network consisting of R1 and R2. Resistor R3, connected in series with the emitter, limits the base current to the desired bias value. Capacitor C1 bypasses AC signals and reduces the degeneration caused by self-bias.

An increase in collector current (caused, say, by an increase in temperature) also increases the base-emitter voltage. This decreases the forward bias on the junction and acts to reduce base current, thus counteracting the original increase. The amount of DC variation depends on the value of R3: a higher value gives better stabilization. The feedback also depends on how constant the base potential can be maintained, and low values of R1 and R2 also improve stability.

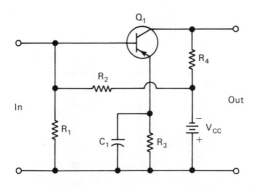

Figure 2-2. Bias with Variable Operating Characteristics

One interesting feature of this circuit is that its operational characteristics can be varied widely by altering the bias resistor network. When R3 is large and R1 and R2 are small, the circuit is equivalent to the high-stability common-base amplifier. When R3 is small and R1 and R2 are very large, the circuit is equivalent to the low-stability, high-gain common-emitter amplifier. The circuit can be operated between these two extremes by selection of R1, R2 and R3.

BOOTSTRAPPED EMITTER FOLLOWER

One reason for using a transistor emitter follower or an FET source follower configuration is for its very high input impedance. However, this input impedance is seriously reduced by the shunting effect of the bias resistors. A technique known as "bootstrapping" is often used in the bias network to maintain the high input impedance of a semiconductor device.

As shown in the emitter follower circuit in Figure 2-3, the base bias resistor has been split in two (R2 and R3 have the same value), and capacitor C2 is connected from this voltage divider to the emitter of transistor Q1. This capacitor applies positive feedback from the output at the emitter to the input of the amplifier. It has a reactance such that almost all the output voltage is coupled back to the junction of R1 and R2.

The output voltage of an emitter follower is slightly less than one. Suppose that the signal applied through C2 to the junction of R2 and R3 is,

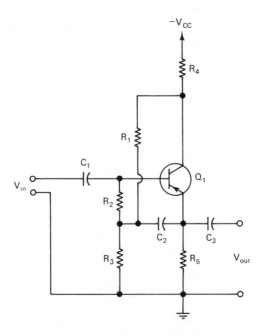

Figure 2-3. Bootstrapped Emitter Follower

say, 0.95 of the input. Since the signal is in phase with V_{IN}, the amount of current through R2 is:

$$I_{R2} = \frac{V_{IN}(1 - 0.95)}{R2} = \frac{V_{IN}}{20\ R2}.$$

If C2 were not present, of course, the current through R2 (and R3) would be:

$$I_{R2} = \frac{V_{IN}}{2\ R2}.$$

So the bootstrapping capacitor causes an effective tenfold increase in the impedance of R2, without affecting the bias conditions of the amplifier. The reactance of C2 at the operating frequency should be at least 0.95 percent of the resistance of R3.

CASCADED TRANSISTOR AMPLIFIER

Figure 2-4 shows how two common-emitter amplifier stages are cascaded, or connected in series, for more gain. The output from NPN transistor Q1 is coupled directly to the base of Q2; emitter resistor R5 ensures that Q2 is biased to match the voltage available at the collector of Q1. The total gain of this simple two-stage amplifier is the sum of the gains in decibels through Q1 and Q2.

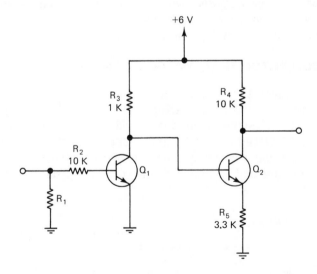

Figure 2-4. Cascaded Transistor Amplifier

COMMON-BASE TRANSISTOR AMPLIFIER

A PNP transistor connected in a simple common-base (or grounded-base) configuration is illustrated in Figure 2-5. Input resistance for this type

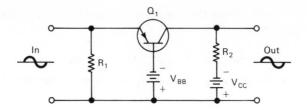

Figure 2-5. Common-Base Transistor Amplifier

of circuit is extremely low, usually about 20-50 ohms, while output impedance is very high, in the range of 1-2 megohms.

In operation, as the signal across R1 goes positive, it acts as an increase in emitter forward bias on Q1, causing an increase in emitter-base current. This small increase causes a similar increase in collector-base current, and the output at the collector of Q1 becomes more positive. A negative-going input signal to the emitter of Q1 reduces the emitter-base current, thus reducing collector-base current and causing the output at the top of R2 to become more negative.

In the grounded-base circuit, the change in emitter current and the change in collector current are nearly the same. Resistance gain between base and collector is high, and voltage gain is normally on the order of 1500 or so. The common-base amplifier, which is similar to a grounded-grid electron tube amplifier, is used where no phase shift is desired or where a very low input impedance and a very high output impedance are wanted.

COMMON-COLLECTOR TRANSISTOR AMPLIFIER

Figure 2-6 illustrates a PNP transistor in a basic common-collector (or grounded-collector) configuration. (Notice that the collector is not actually grounded, but an AC signal ground is provided through capacitor C1.) The input signal is developed across R1 and is applied to the base of Q1. Input impedance is very high, normally from 150,000 to 600,000 ohms. Output is taken off the emitter and its impedance is very low, usually on the order of 100 to 1000 ohms.

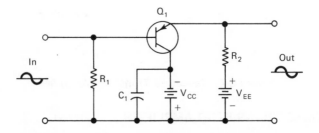

Figure 2-6. Common-Collector Transistor Amplifier

When the input signal swings positive, it effectively increases reverse bias on the base-collector junction, causing a decrease in current through the emitter-base junction. This in turn produces a decrease in the output voltage developed across R2; so the output potential at the collector becomes more positive. Similarly, when the input signal swings negative, it causes an increase in the voltage across R2, and the output goes negative.

Voltage gain for the grounded-collector amplifier is always less than unity. Current gain is high, so power gain is also high, although not as high as for either a grounded-emitter or grounded-base configuration. The common-collector amplifier is similar to an electron tube connected in a grounded-plate configuration, and is usually used for impedance matching. In addition, the common-collector amplifier can actually pass a signal in either direction.

This type of circuit is also called an *emitter follower*.

COMMON-DRAIN FET

The basic common-drain field-effect transistor amplifier illustrated in Figure 2-7 offers very high input impedance, low output impedance, and a voltage gain less than unity. This circuit, also known as a *source follower*, does not invert the output. It provides low distortion and high power gain.

Q1 is an N-channel FET, and R1 is the gate resistor. R2 is the load resistor in this very much simplified circuit.

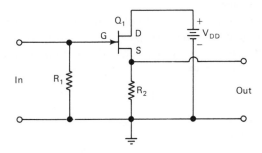

Figure 2-7. Common-Drain FET

COMMON-EMITTER TRANSISTOR AMPLIFIER

Figure 2-8 illustrates the simplest form of a common-emitter, or grounded-emitter, amplifier configuration. Input impedance of this type of circuit is low to medium, usually in the range of 1000 to 2000 ohms. Output impedance is on the order of 50,000 ohms. The input and output signals are 180° out of phase with one another.

When a signal is applied across base resistor R1, the base current changes according to the signal. When the input is positive to the base of

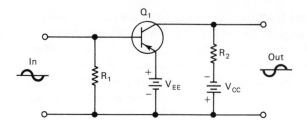

Figure 2-8. Common-Emitter Transistor Amplifier

PNP transistor Q1, it is opposite in polarity to the emitter-base bias provided by V_{EE}, and the base-emitter current is reduced. This decrease in base-emitter current causes a similar decrease in emitter-collector current, with the result that the voltage at the collector of Q1 becomes more negative.

When a negative signal is applied to the base of Q1, emitter-base current increases, causing emitter-collector current to increase also. This increase in collector current allows more voltage to develop across load resistor R2, with the result that the output at the collector of Q1 becomes more positive.

As an amplifier, the common-emitter circuit yields a large voltage gain, on the order of 1500 or so, and also provides large current gain. It has the highest power gain of the three common amplifier configurations for transistors. The common-emitter amplifier is similar to the grounded-cathode electron tube amplifier.

Common-emitter direct current gain is one of the parameters most often listed in the specifications for a transistor. It is referred to by the symbol α', β, or h_{FE}, and it is given by the ratio of the change in collector current to the change in base current:

$$h_{FE} = \frac{I_C}{I_B}$$

COMMON-GATE FET

In the basic common-gate FET amplifier, illustrated by Figure 2-9, the input signal is fed into the source and taken off the drain. This type of circuit is valuable for impedance matching, since its input impedance is very low and its output impedance is high. It has low voltage gain.

COMMON-SOURCE FET

Figure 2-10 shows an N-channel field-effect transistor connected in the most commonly used configuration for this device, a common-source arrangement. Input is fed to the gate lead of the FET, and the output is taken off the drain. R1 is the gate resistor, while R2 represents the load resistor. The common-source amplifier has high input impedance, medium to high output impedance, and a voltage gain greater than one. Output is inverted from input.

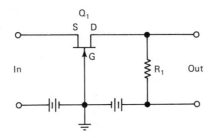

Figure 2-9. Common-Gate FET

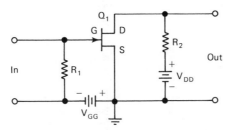

Figure 2-10. Common-Source FET

CURRENT REGULATOR DIODE APPLICATIONS

Current regulator diodes in ratings from 20 microamperes to 517 milliamperes are available from Siliconix in their NKL, NKM and MKO series. They are similar in operation to Zener diodes. Figure 2-11 shows some applications for these current regulator diodes. They can be used as collector drain loads for providing constant current in timing circuits, as a logic circuit pull-up current source and for furnishing emitter source biasing. (© *Siliconix incorporated.*)

DARLINGTON CIRCUIT

A Darlington circuit uses two direct-coupled NPN (or two PNP) transistors, an emitter-follower (common-collector) driving another emitter-follower, as shown in Figure 2-12. The circuit provides less than unity voltage gain, but power gain of almost 200 times. Input impedance varies from 2000 ohms to 200,000 ohms and output load impedance from 10 ohms to 1000 ohms, depending on the values of the resistors and characteristics of the transistors.

DEGENERATIVE COMMON-SOURCE FET AMPLIFIER

In the common-source FET amplifier stage shown in Figure 2-13, the source resistor has been split into resistors R2 and R3. Decoupling capacitor

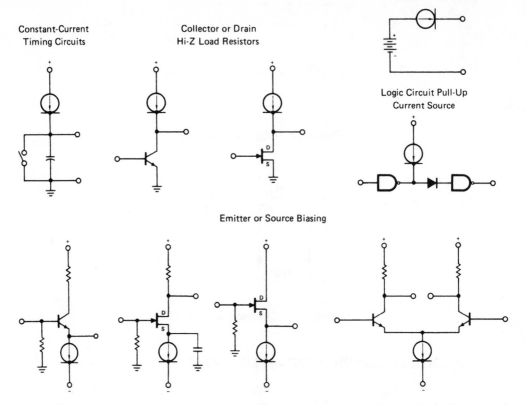

Figure 2-11. Current Regulator Diode Applications

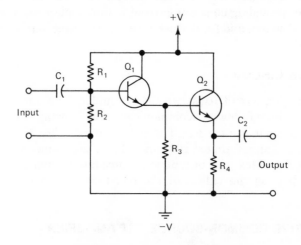

Figure 2-12. Darlington Circuit

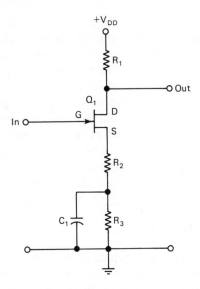

Figure 2-13. Degenerative Common-Source FET Amplifier

C1 and resistor R3 act as in a self-biased, common-source amplifier. This means that part of the signal appears at the drain, as in a normal common-source amplifier, and part of the signal appears at the source, as in a common-drain amplifier.

Gain with this configuration is less than for a normal common-source amplifier, but stability of the circuit is increased. If R1=R2, the outputs at the source and drain will be equal in amplitude and opposite in phase.

EMITTER BIAS

Emitter bias of a transistor is a form of self-bias. As shown in Figure 2-14, emitter resistor R3 and capacitor C1 are connected between the emitter of common-emitter amplifier Q1 and ground.

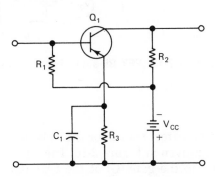

Figure 2-14. Emitter Bias

In operation, C1 and R3 act much like the capacitor and resistor in a cathode-biased vacuum tube. The value of R3 is usually 1/10 or less of the value of R1, so almost all collector current flows through R3. C1 provides an AC signal path; its value depends on the lowest frequency to be amplified. DC current through R3 makes the emitter negative with respect to ground. The base is, of course, also negative with respect to ground, and the base-emitter voltage is approximately equal to the difference between the supply voltage drop across R1 and the small reverse voltage dropped across R3.

If a temperature increase tends to cause the collector current to increase, the voltage across R3 also increases, thus decreasing the base-emitter voltage. This decrease in forward bias of the base-emitter circuit decreases collector current, thus counteracting the increase caused by temperature.

FET BIAS AND PARAMETERS

Figure 2-15 shows the basic principles of field-effect transistor circuits. The N-channel FET is biased much like an electron tube. External power supplies are connected so that the drain is biased positive with respect to the source, and drain current (I_D) flows. In order for the gate to control current flow, it is biased negative with respect to the source. V_{GS} is the value of gate bias, while V_{DS} is the voltage dropped from source to drain. The P-channel FET is biased with the drain negative and the gate positive with respect to the source. (© *Siliconix incorporated.*)

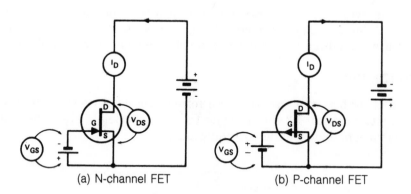

(a) N-channel FET   (b) P-channel FET

Figure 2-15. FET Bias and Parameters

FET INPUT STAGE

An N-channel field-effect transistor (FET) is used as a common source amplifier in the circuit given in Figure 2-16. The input signal is fed through coupling capacitor C1 to the gate (G) of the FET (Q). The FET is reverse-biased by resistor R2 in series with the FET source(S), which makes the gate negative with respect to the source. The input signal varies FET drain (D) current through R3. The output signal is fed through C3 to the next stage.

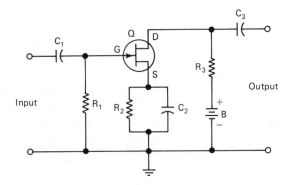

Figure 2-16. FET Input Stage

The input resistance is high since the FET gate draws no current as long as it is reverse biased. Capacitor C2 across R2 grounds Q's source for AC. If C2 is omitted from the circuit, degenerative feedback will reduce gain but will also reduce distortion. (In the diagram, battery B represents the drain voltage source.)

FIXED AND SELF-BIAS

Figure 2-17 illustrates how a voltage divider is used to obtain both fixed bias and self-bias on common-emitter amplifier Q1. The voltage divider consists of base resistor R1 and self-biasing resistor R3.

If a higher temperature tends to increase collector current, the voltage at the collector of Q1 becomes more positive. This increased positive potential is felt through R3, and acts to reduce the forward emitter-base bias, thus reducing current through this region. Reduced emitter-base current, in turn, acts to reduce collector current and counteract the original increase. As with self-bias, negative signal feedback in this circuit decreases the gain of the amplifier.

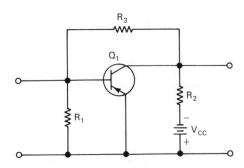

Figure 2-17. Fixed and Self-Bias

FIXED BIAS

Figure 2-18 illustrates a PNP transistor connected as a basic common-emitter amplifier using fixed bias. In fixed bias, base resistor R1 is connected directly to the supply voltage.

The value of R1 is determined by the desired base current (I_b), the supply voltage (V_{cc}) and the base and emitter resistances of the transistor ($r_b + r_e$). Since R1 is usually a very large resistance, in the range of several hundred thousand ohms, while r_b and r_e are comparatively small, the internal resistances of Q1 can, for all practical purposes, be ignored when computing a value for base resistor R1. Thus:

$$R1 \simeq \frac{V_{cc}}{I_b}.$$

Because of variations in transistors and the change in their electrical characteristics with varying temperature, it may be difficult to maintain close tolerances of base current with fixed bias alone.

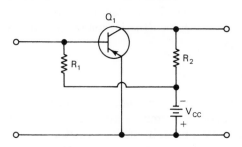

Figure 2-18. Fixed Bias

INVERTING OPERATIONAL AMPLIFIER

In Figure 2-19, an operational amplifier is connected with feedback from the output to the inverting (−) input. If the ratio of R2/R1 is small compared to the amplifier open-loop gain, the output of this configuration is given by:

$$V_{OUT} = V_{IN} \times \frac{R2}{R1}.$$

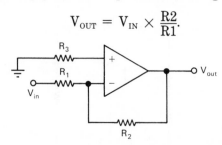

Figure 2-19. Inverting Operational Amplifier

To minimize offset error, resistor R3 should be selected to be equal to the parallel combination of R1 and R2. In addition, the offset voltage at the output is equal to the offset voltage at the input multiplied by the closed-loop gain.

NONINVERTING OP AMP

Figure 2-20 shows the basic configuration for a noninverting operational amplifier. The input is applied to the noninverting (+) input. Feedback is applied to the inverting input through resistor R2.

For this circuit,

$$V_{OUT} = \left(\frac{R1 + R2}{R1}\right) V_{IN}.$$

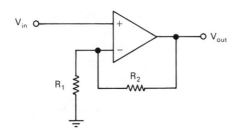

Figure 2-20. Noninverting Op Amp

SELF-BIAS

Self-bias of a transistor, as shown in the common-emitter configuration in Figure 2-21, generally provides more stable operation than does fixed bias. In self-bias, base resistor R1 is connected to the collector of Q1 rather than directly to the supply voltage V_{cc}. The value of the base resistor is very large

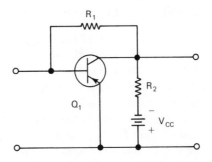

Figure 2-21. Self-Bias

compared to the internal resistances of Q1, so the collector voltage and the desired base current can be used to calculate a value for R1:

$$R1 \simeq \frac{V_c}{I_b}.$$

In operation, if collector current increases because of a rise in temperature, the voltage drop across load resistor R2 also increases, thus decreasing V_c, the voltage at the collector. This decrease causes base current I_b to decrease, which in turn *decreases* collector current. Self-bias thus produces more stable operation, but negative feedback of the signal also reduces the effective gain of the amplifier stage.

STABILIZED BIAS

The common-emitter transistor amplifier in Figure 2-22 uses a divided DC return and a capacitor to obtain both self-bias and fixed bias. Capacitor C1 bypasses all AC signal variations, helping to reduce degeneration caused by the self-bias. The value of R1 is usually 1/5 to 1/10 that of R2. This type of bias allows stabilization with a minimum amount of degeneration.

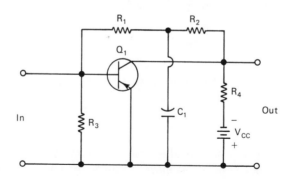

Figure 2-22. Stabilized Bias

UNIJUNCTION TRANSISTOR AS PULSE GENERATOR

Figure 2-23 illustrates the operation of the unijunction transistor (UJT) in a pulse generating circuit. At the outset, the emitter is reverse-biased and a steady current flows from base B2 to B1. As capacitor C1 charges through R3, the bias on the emitter becomes more positive until it exactly counterbalances the reverse bias the emitter is picking off from the base region.

As the positive voltage across C1 continues to rise, the emitter becomes forward-biased, and emitter current flows through the more negative portion of the base (B2). This current effectively reduces the resistance of that portion of the base, allowing C1 to discharge quickly through the emitter-B2 junction and R2. With C1 discharged, the emitter is once again reverse-biased, and the cycle begins again, producing a sawtooth waveform at the output.

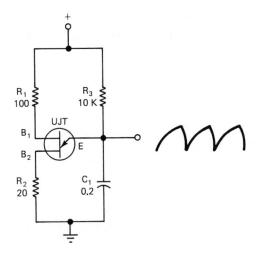

Figure 2-23. Unijunction Transistor as Pulse Generator

VARACTOR-TUNED RESONANT CIRCUITS

One varactor diode can be shunted across a coil to form a tunable resonant circuit. Tuning is accomplished by varying a DC voltage which is applied to the varactor to vary its capacitance. However, a single varactor can be affected by a signal voltage whose peak exceeds the DC bias applied to the varactor. By connecting two varactor diodes in series opposing across the coil, the DC bias is not affected by the RF signal voltage. Figure 2-24 shows how two circuits can be tuned with one potentiometer. A second potentiometer is used for presetting one of the resonant circuits to obtain proper tracking. This tuning technique can be applied to a TRF receiver for tuning the RF amplifier and detector, a superheterodyne receiver for tuning the mixer input and the local oscillator, an IF amplifier that is tunable (to

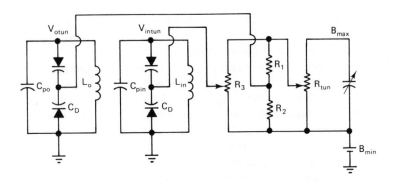

Figure 2-24. Varactor-Tuned Resonant Circuits

enable using a fixed tuned frequency converter ahead of the amplifier), an RF signal generator for tuning the oscillator and an RF buffer stage, and in many other devices.

The diagram shows an application for the use of a potentiometer as the main tuning control. DC bias is applied to the varactors connected across coils L_o and L_{in}. The bias on the varactors across L_{in} can be preset with potentiometer R3. The bias on the varactors across L_o is preset by the fixed voltage divider (R1 and R2). The tuning voltage source is represented in the diagram by Bmax and the minimum bias is represented by battery Bmin. Note that the voltage source reverse-biases the varactors (cathode is made positive with respect to anode). Typically, the tuning voltage is varied from 1.8 to 7.5 volts. Battery Bmin would be 1.8 volts and Bmax would be 7.5 volts. Typically, the diode capacitance would be varied form 8 to 25 pF. When a 0.3 microhenry coil is used, varying the capacitance of two series-connected varactors through the 7 to 12 pF range would enable tuning through the 88-108 MHz FM broadcast band. When two varactors are connected in series, their effective capacitance is half of the capacitance of each varactor.

The frequency range can be extended by using varactors having a wider capacitance range. And the frequency range can be lowered by increasing the inductance of the coils or by shunting the varactors with fixed capacitors. The voltage source should be a battery or a well-regulated AC-to-DC power supply.

VARISTOR PROTECTION

When an inductive load is switched off by a transistor, a high surge of reverse bias is applied which can damage the transistor if its rating is not high enough. As shown in Figure 2-25, connecting a varistor in parallel with the emitter-collector circuit shunts the sudden high-voltage transient away from the transistor, thus protecting it. (*Courtesy of General Electric Company.*)

ZENER DIODE VOLTAGE REGULATION

Figure 2-26 shows the most common use of the Zener diode, in a basic voltage regulating circuit. The diode is reverse-biased by the supply voltage and is operated close to its Zener point.

If the voltage from the supply increases, current through Zener diode D1 and the load (represented by resistor R2) will also increase. This increased current causes a drop in the resistance of D1, which in turn causes the voltage at the junction of R1 and the cathode of D1 to remain constant. So, although the increased current is seen in an increased voltage drop across R1, current through the load remains constant.

In the case of a varying load, attempting to increase the current through the load reduces the reverse current through D1, while decreasing the load current increases the current through the Zener diode. The voltage across D1 (and hence, across the load) remains constant.

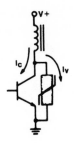

Figure 2-25. Varistor Protection

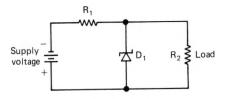

Figure 2-26. Zener Diode Voltage Regulation

CHAPTER 3

Audio Amplifiers

INTRODUCTION

Although the audio range extends from 20 Hz to 20 kHz, many people cannot hear all sounds in that frequency range. But some people who cannot hear certain sounds can feel the effects of those sounds. The object of an audio amplifier is to faithfully reproduce all the sounds in the audio range. (Or in the desired portion of the range. For example, an audio amplifier used in telephone circuitry must amplify only up to about 4 kHz, although audio sounds, including human speech, go higher.)

Although perfect reproduction of sound has seldom been achieved, the first goal of an audio amplifier is *fidelity*—the reproduction of sound without distortion. At the same time, the output of an AF amplifier must often be at a fairly high power level—as much as 30 watts or more—so the gain of each amplifier stage must be at its maximum.

One method used to help balance high gain with low distortion is to feed back a portion of the amplified signal to an earlier stage in the amplifier. Positive, or regenerative, feedback is a signal that is fed back *in phase* with a signal in a previous stage. Positive feedback increase the gain of an amplifier, but it also tends to cause distortion and even oscillation. When, on the other hand, part of an amplified signal is fed back *out of phase*, it is called negative, or degenerative, feedback. The use of negative feedback in an amplifier decreases its gain, but increases stability, reduces distortion and drift, and can even improve frequency response.

Another consideration in amplifier circuitry is the *class* of operation. In a class A amplifier, the positive and negative swings of the input signal drive the amplifying device above and below a selected operating point, but the device is always turned on. The class A amplifier gives low distortion, and is especially useful when power consumption is not a factor.

In class B amplification, the driving element is turned on for only half of each cycle of input (as in, for example, the push-pull amplifier). Class B operation of a semiconductor device is more efficient than class A because the device is turned on for only half of each cycle of input.

For audio amplifiers, where the faithful reproduction of an input signal is of greatest importance, most devices are operated class A, or class AB in push-pull. True class B push-pull amplification usually has too much distortion at the crossover point (where the input drive causes one transistor or FET to cut off and the other to turn on) to be used in audio amplifiers. In class AB, however, a small forward bias is used to minimize the nonlinearity of the crossover point, while maintaining the high efficiency of the class B amplifier stage.

Why use class AB or class B at all? The advantages are that class B or AB amplifiers require low standby power, are highly efficient, and yield high power from relatively low-power semiconductor devices.

AC LINE-OPERATED AUDIO AMPLIFIER

Direct operation from the AC power line (without a power transformer) is a feature of the audio amplifier circuit given in Figure 3-1. One side of the AC line is fed through an on-off switch, a 1-ampere fuse, a diode rectifier, a voltage-dropping resistor, and a filter resistor. At the junction of the two resistors a filter capacitor is connected and another filter capacitor is connected to the load end of the filter resistor. The DC output across this capacitor is fed to the 2N3584 output transistor through the primary of the output transformer. The DC output is also fed through a voltage-dropping resistor to the collector load resistor of the 40231 input transistor. This amplifier will deliver 5 watts to an 8-ohm speaker. Its frequency response is 40-4000 Hz within -3 dB. This amplifier lends itself to application in radio receivers, inexpensive phonographs and other devices in which the inclusion of a power transformer would add too much to the cost. (*Courtesy of RCA Solid State.*)

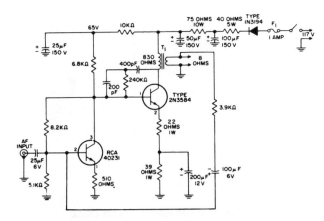

Figure 3-1. AC Line-Operated Audio Amplifier

AUTO STEREO TAPE DECK

The circuit of the RCA 12R902 auto tape player is given in Figure 3-2. Outputs of the left and right tape heads are amplified by ICI (LM5152) whose left and right channel outputs are fed through a dual 10,000-ohm volume control to the inputs of IC2 and IC3 (both are MPC100H or MPC10204) for further amplification. The outputs of IC2 and IC3 are fed to an external 3-8 ohm speaker for each stereo channel. The DC operating power is derived from the electrical system of the vehicle in which the tape player is installed. (*Courtesy of RCA Solid State.*)

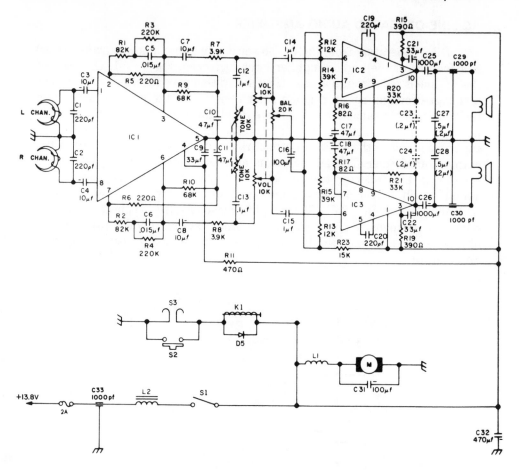

Figure 3-2. Audio Stereo Tape Deck

BASS/TREBLE EQUALIZER

A 325EQ equalizer amplifier application is given in Figure 3-3. By adjusting the 1000-ohm linear potentiometer at the left, the bass response can be varied from +18 dB to −18 dB. By adjusting the pot to the right of the bass control, the treble response can also be adjusted over the same range. A ±24 volt DC power supply is required. The output level is controllable with the 100-ohm pot at pin 4. (*Courtesy of Opamp Labs Inc.*)

BAXENDALL TONE CONTROL

With the Baxendall tone control circuit, shown in Figure 3-4, the amount of feedback at both low and high frequencies can be varied, allowing control of the tone of an audio amplifier.

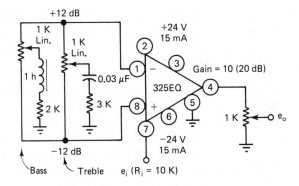

Figure 3-3. Bass/Treble Equalizer

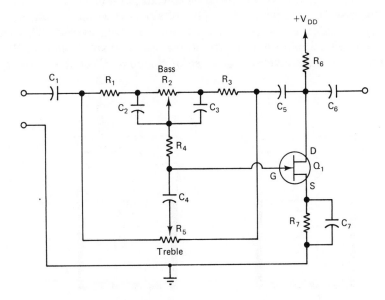

Figure 3-4. Baxendall Tone Control

Input to the amplifier stage enters through capacitor C1 and resistor R1, passes through potentiometer R2 and resistor R4, and is applied to the gate of field-effect transistor Q1. Negative feedback is applied from the drain of Q1 through capacitor C5, resistor R5, and capacitor C4. Since the reactance of C5 decreases as frequency increases, R5 provides higher-frequency treble control by picking off more or less of the feedback; moving the wiper arm of R5 toward C5 picks off more high-frequency feedback and less signal.

Another path for feedback is through resistor R3. This is used to control the bass response of the amplifier. Bass control is provided by resistors R1, R2 and R3 and capacitors C2 and C3. Moving the wiper arm of R2 toward resistor R1 decreases feedback and increases the amount of signal, effectively boosting the amplification of low-frequency signals.

BRIDGED AUDIO POWER AMPLIFIER

A pair of Sprague ULX-3701Z/TDA2002 ICs connected as shown in Figure 3-5 can deliver 15 watts into a 4-ohm speaker. Each of these ICs is packaged as a 5-lead JEDEC style TO-220 plastic package. Input sensitivity varies from 15 millivolts for 50 milliwatts output to 60 millivolts for full-rated output. The input signal from the radio detector or an audio preamplifier is fed into pin 1 of the IC at the left. Pin 2 of the IC at the right is the input signal source for that IC. It gets its input signal from the outputs of both ICs through a 220-ohm resistor in each leg. Balance is controlled with the 100,000-ohm potentiometer that is connected to pin 5 of both ICs through a 1-megohm resistor. The nominal supply voltage is rated at 14 volts which is commonly available in autos with a 12-volt electrical system. This circuit should have wide application in automotive stereo systems, mobile PA systems and in CB transceivers. (*Courtesy of Sprague Electric Company.*)

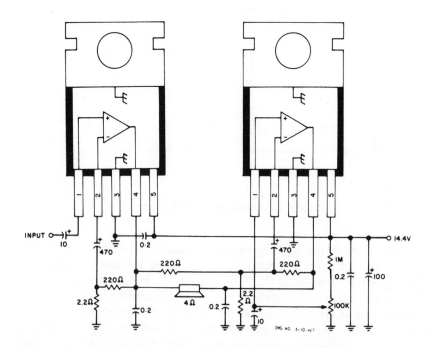

Figure 3-5. Bridged Audio Power Amplifier

COMPLEMENTARY SYMMETRY AMPLIFIER

The current paths in a PNP and an NPN transistor are opposite in polarity. This feature is put to use in the complementary symmetry amplifier configuration, a basic example of which is shown in Figure 3-6.

This amplifier uses a pair of transistors connected as common-collector amplifiers, and operates in a manner similar to the push-pull amplifier: only one transistor is on at a time. In the complementary symmetry amplifier, however, transistor Q1 is a PNP device, while Q2 is an NPN transistor.

When the input signal swing is negative, the base of Q1 is forward-biased and Q1 conducts. At the same time, Q2 is cut off. The output across load resistor R4 is also negative at this time, so the output and input are in phase with each other.

When the input signal swing is positive, Q1 is cut off and Q2 conducts. Again, the output across R4 is the same polarity as the input signal.

This type of circuit has a distinct advantage over the conventional push-pull amplifier because by using complementary symmetry no center-tapped transformer is necessary.

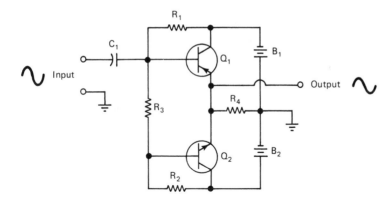

Figure 3-6. Complementary Symmetry Amplifier

DIFFERENTIAL OPERATIONAL POWER AMPLIFIER

An op-amp system capable of delivering 50 watts RMS to an 8-ohm speaker system or to a servo motor is depicted in Figure 3-7. The output current is rated at 10 amperes maximum. Input signal level and output voltage swing at ±30 volts. Open loop plate gain is 800. As can be seen in the diagram, two PNP and three NPN power transistors and four diodes are used in addition to a 4009 op amp. Since the open loop output impedance is 0.1 ohm, the voltage regulation is excellent. (*Courtesy of Opamp Labs Inc.*)

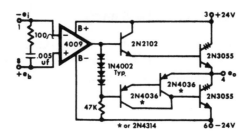

Figure 3-7. Differential Operational Power Amplifier

DIRECT-COUPLED AUDIO AMPLIFIER WITH FEEDBACK

Negative feedback is provided from the collector of Q2 to the emitter of Q1 in the direct-coupled amplifier circuit given in Figure 3-8.

This negative feedback lowers the output impedance of Q2, the overall gain and the distortion. General purpose transistors may be used. The amount of feedback may be reduced by using a higher-value resistor in place of the upper 35K resistor, using a lower-value capacitor in lieu of the 0.05 μF capacitor shown. By applying the feedback to the emitter of Q1, and leaving the 1200-ohm Q1 emitter resistor unbypassed, the feedback signal is effectively fed to the Q1 base, whose input resistance is high.

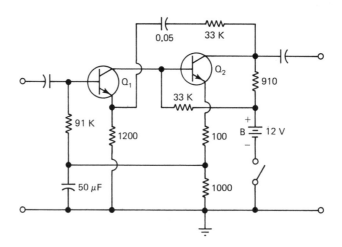

Figure 3-8. Direct-Coupled Audio Amplifier with Feedback

DIRECT-COUPLED AUDIO AMPLIFIER WITH LOW IMPEDANCE OUTPUT

Two NPN transistors are direct-coupled in the circuit given in Figure 3-9. This amplifier has a rated gain of 40 dB. Q2 is utilized as an emitter-follower to provide a low impedance output.

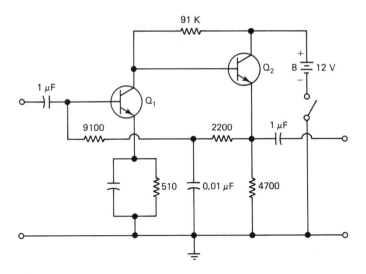

Figure 3-9. Direct-Coupled Audio Amplifier with Low-Impedance Output

FIVE-WATT IC AF AMPLIFIER

A single TBA800 IC is used in the amplifier circuit given in Figure 3-10. It is capable of delivering 4-5 watts into a 16-ohm speaker when operated from a 24-volt DC source. The gain of the amplifier is rated at 80 dB. Capacitor Cx may be a 100-μF, 25-volt electrolytic type. Try various values for Rx starting with 1 ohm.

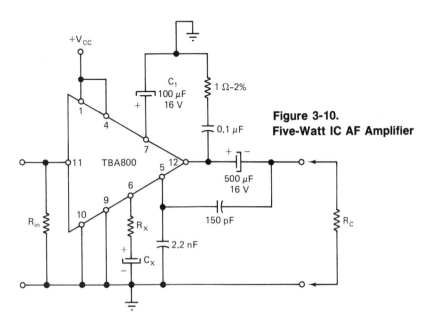

**Figure 3-10.
Five-Watt IC AF Amplifier**

HIGH FIDELITY 40-WATT VMOS AUDIO AMPLIFIER

A pair of three paralleled 2N6658 VMOS transistors is used in the output stage of the 40-watt high fidelity audio amplifier whose circuit is given in Figure 3-11. Negative feedback (22 dB) from the junction of the sources of Q11-Q13 and the drains of Q8-Q10 to the base of Q2 extends frequency response to 400 kHz and minimizes transient intermodulation distortion. Matching of the N-channel VMOS devices in these quasi-complementary drains is achieved by providing local feedback from the drains to the gates of Q14-Q16 through R15. The gates are then driven with a modulated current source.

This amplifier circuit utilizes current regulating diodes (D1, D3, D6, D7) to maintain constant current in critical circuits. Zener diodes (D8, D9) are used to stabilize and limit output current and dissipation by limiting gate enhancement.

An RF filter at the input of the amplifier minimizes pickup of RF signals that are within the frequency range of the amplifier. Adjusting potentiometer R14 enables exact matching of the positive and negative output waveform during class AB operation. (© *Siliconix incorporated.*)

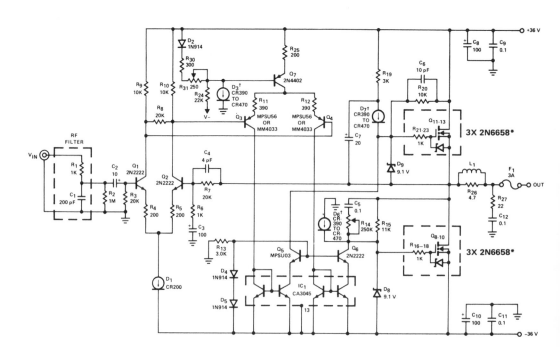

Figure 3-11. High Fidelity 40-Watt VMOS Audio Amplifier

HIGH GAIN AUDIO AMPLIFIER

The circuit of a 5-watt audio amplifier with 44 dB of gain is given in Figure 3-12. A Plessey SL415A 10-pin dual-in-line IC is used in this circuit. This IC contains a preamplifier with 24 dB of voltage gain and a main amplifier that provides an additional 20 dB of voltage gain. Peak output power is 5.5 watts when 24 volts DC is used to power the IC, dropping to 4 watts when used with a 20-volt supply. (*Copyright Plessey Semiconductors.*)

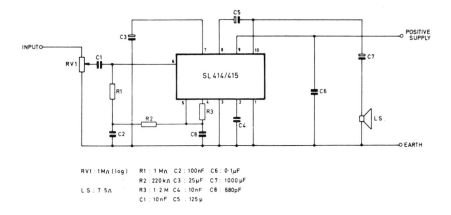

RV1 : 1MΩ (log)	R1 : 1 MΩ C2 : 100nF C6 : 0·1μF	
	R2 : 220kΩ C3 : 25μF C7 : 1000 μF	
LS : 7·5Ω	R3 : 1·2 M C4 : 10nF C8 : 680pF	
	C1 : 10nF C5 : 125μ	

Figure 3-12. High Gain Audio Amplifier

LOW POWER AUDIO AMPLIFIER

A single 8-pin dual-in-line AC is used in the circuit given in Figure 3-13. As a low power audio amplifier, it will function with supply voltages

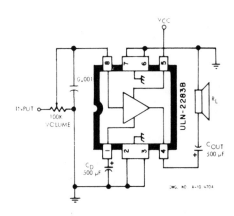

Figure 3-13. Low Power Audio Amplifier

as low as 2 volts. (The Sprague ULN-2283B IC is rated for operation over the 3-15 volt supply range.) The voltage gain is 19 dB when a 3-volt supply is used, rising to 46 dB when 15 volts is supplied. Audio output power is rated at 250-350 milliwatts into an 8-ohm speaker when 3 volts DC is supplied. Output rises to 800-1200 milliwatts into a 60-ohm speaker when supply is 15 volts. As can be seen in the diagram, only four external components are required: a 100,000-ohm volume control, an 0.001-uF frequency compensation capacitor, a 500-μF electrolytic bypass capacitor and a 500-μF electrolytic capacitor that isolates the DC from the speaker. (*Courtesy of Sprague Electric Company.*)

MOBILE POWER SPEAKER

Back in the late 1920s, the RCA 104 and 105 power speakers and the Kolster power speaker were widely used at dance halls, picnic grounds, in front of stores, etc. to provide amplified music and talk programs. Although the output power of the power speaker was around 1 watt, they sounded quite loud.

Today, some stereo amplifiers deliver upward of 100 watts to highly efficient speakers. In addition, there are amplified mobile speakers that are used with mobile PA systems and to reinforce the output of mobile radio transceivers.

In 1947, Ernest A. Dahl, who was then radio engineer for the Rock Island Lines, designed an amplified speaker for use in railway coaches to distribute background music. These amplified speakers were recessed in the ceiling of a railway coach and contained a power amplifier utilizing four 35L6GT tubes in push-pull parallel. The power source was the 32-volt car-lighting battery. The tube heaters were connected in parallel across the 32-volt input. The plate voltage for the tubes was also the same 32-volt source which was adequate for producing sound at the required level. Tubes had to be used because transistors were not invented until that year.

The circuit of a modern solid state power speaker is given in Figure 3-14. It utilizes a pair of 40250 NPN power transistors in push-pull. With a 750-milliwatt input signal at the primary of T1, the amplifier will deliver 10 watts to a 3.2-ohm speaker. This amplifier is designed to operate from a 12-volt vehicular electrical system which delivers a nominal 13.8 volts. Input current is 5 amperes maximum. (*Courtesy of Hallicrafters.*)

MONO PHONOGRAPH AMPLIFIER

A single Sprague ULN-2281B IC is used in the 4-watt monaural phonograph amplifier circuit given in Figure 3-15. The output of a ceramic phonograph pickup is fed through a 75,000-ohm series resistor and a 25,000-ohm volume control. The signal is fed from the potentiometer and a 0.01-μF capacitor to pin 2 of the 14-pin dual-in-line IC. Pin 8 is connected through a 250-μF capacitor to an 8-ohm or 16-ohm speaker whose common lead is

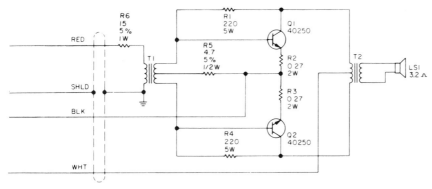

NOTE ALL RESISTORS ARE IN OHMS

Figure 3-14. Mobile Power Speaker

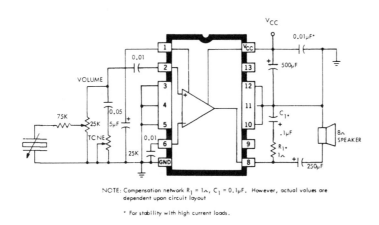

NOTE: Compensation network $R_1 = 1\Lambda$, $C_1 = 0.1\mu F$. However, actual values are dependent upon circuit layout

* For stability with high current loads.

Figure 3-15. Mono Phonograph Amplifier

grounded to pins 10, 11 and 12. Shunted across the amplifier output is a frequency equalizer consisting of a 0.1-µF capacitor in series with a 1-ohm resistor. (*Courtesy of Sprague Electric Company.*)

POWER MICROPHONE

A power microphone contains a preamplifier as shown in Figure 3-16. The microphone has an output that is essentially 6 dB above that normally obtained with a conventional carbon-type microphone. This permits the operator to speak in a normal voice at a comfortable distance (one to two feet). It may be found in some installations with high ambient background noise that the sensitivity of the microphone is excessive. In these instances, the

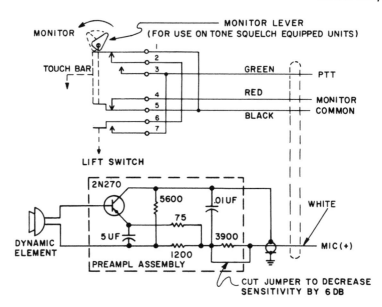

Figure 3-16. Power Microphone

output level of the microphone may be reduced by clipping the jumper provided for this purpose.

The DC voltage necessary for operation of the transistor amplifier may
be supplied by the microphone input circuit of the associated transmitter.
This voltage is the same as that required to operate a standard carbon
microphone. The microphone shown is directly interchangeable with carbon
microphone-equipped units. (*Courtesy of Hallicrafters.*)

PUSH-PULL AMPLIFIER

Many audio power amplifiers use a push-pull configuration as the final
output stage. Figure 3-17 shows a basic push-pull amplifier stage made up
of two PNP transistors and a center-tapped transformer T2.

When the input signal swings positive, the signal at the center tap of
input transformer T1 is negative with respect to the top of the secondary
winding and positive with respect to the bottom. Thus, transistor Q2 is turned
on and conducts, while Q1 is cut off. On the opposite swing of the input
signal, Q1 conducts and Q2 is cut off. Each half-cycle is developed across
half of the primary of T2, and the entire sine wave is available at the
secondary of this transformer.

With no bias applied, each transistor operates class B. One problem
with push-pull amplifiers is crossover distortion, which occurs at the point
when one transistor is cut off and the other has not yet started to conduct.
So resistor R2 is present to slightly forward-bias both base-emitter junctions,
keeping one transistor on slightly at the crossover point until the other has

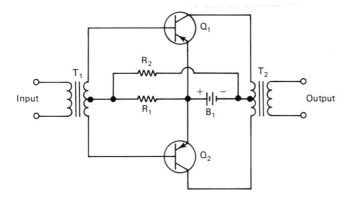

Figure 3-17. Push-Pull Amplifier

started to conduct. This biasing significantly reduces crossover distortion; with it, the amplifier is actually operated in class AB.

For simplicity, the input to the amplifier is shown developed across a center-tapped transformer. This is not necessary if another means of alternating the phases of the base signals is used.

RC-COUPLED AMPLIFIER

One of the most common types of coupling between amplifier stages is RC (resistor-capacitor) coupling. In the two-stage amplifier shown in Figure 3-18, the output signal from Q1 is developed across resistor R2 and coupled through capacitor C1 to the base of Q2, the next stage; thus, R2 and C1 form the RC coupling network. To couple audio frequencies, C1 must be relatively large, on the order of 10 microfarads or so.

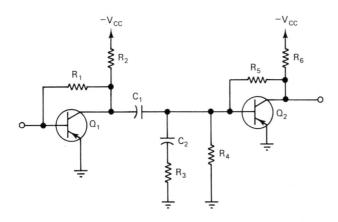

Figure 3-18. RC Coupled Amplifier

Capacitor C2 and resistor R3 form a low-frequency compensation network. C2 is very small (on the order of 1 or 2 microfarads) and shunts only higher frequencies to ground. This has the effect of "boosting" lower frequencies.

SINGLE-ENDED VMOS AF POWER AMPLIFIER

A single VN66AF VMOS transistor is used in the AF power output stage of the audio amplifier circuit given in Figure 3-19. This amplifier has a rated output power of 4 watts into an 8-ohm speaker within the 100-15,000 Hz frequency range. The drain circuit is coupled to the amplifier through an output transformer with a 24-ohm primary and an 8-ohm secondary with a 4-ohm tap. This amplifier circuit can be used as the audio amplifier output stage of a portable phonograph, radio receiver, TV set or as an independent low-power amplifier. (© *Siliconix incorporated.*)

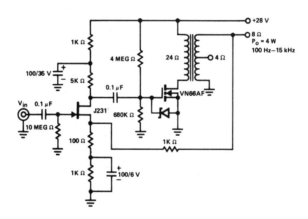

Figure 3-19. Single-Ended VMOS AF Power Amplifier

SINGLE IC STEREO AUDIO AMPLIFIER

A single Sprague ULN2278B IC is used in the stereo amplifier circuit given in Figure 3-20. This 14-pin dual-in-line IC contains both the left and right channel amplifiers and provides up to 60 dB of channel separation. As can be seen in the diagram, only four capacitors and five resistors are required for each channel; and one external capacitor is required for common use by both channels. Audio output power is rated at 2 watts per channel when supply voltage is 9 volts and 2.5 watts when an 18-volt supply is used. When 8-ohm speakers are used (represented by the R_L in the diagram), output power ranges from 0.5 watt to 3 watts with supply voltage in the 8 to 20 volt range. Output power to 16-ohm speakers ranges from 0.5 watt with 10 volt supply to 2 watts with 22.5-volt supply. (*Courtesy of Sprague Electric Company.*)

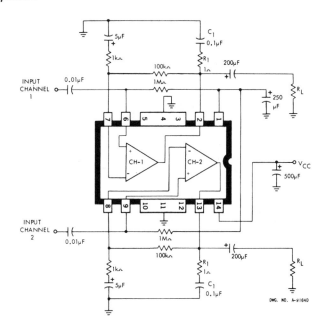

NOTES:
1. Compensation network: R_1, C_1 values are dependent upon circuit layout.
2. When an unregulated supply voltage is used, the actual voltage present at pin 14 during full signal conditions should not drop below the nominal supply voltage level if full power output is to be maintained.
3. Closed loop gain should be limited to 30dB min. to 60dB max. to maintain stable circuit operation.

Figure 3-20. Single IC Stereo Audio Amplifier

SINGLE IC 8-WATT AUDIO POWER AMPLIFIER

The circuit of an 8-watt audio power amplifier utilizing a single 5-pin IC in a plastic package is given in Figure 3-21. The LM2002AT IC provides 40 dB of gain and, when operated from a 16-volt DC source with a 4-ohm speaker as the load, total harmonic distortion is rated at 10 percent. (*Copyright National Semiconductor Corporation.*)

SOUND-ON-FILM PREAMPLIFIER

A solid-state preamplifier for use in a movie projector is shown in Figure 3-22. It amplifies the output of a photocell used for picking off the sound of an optical soundtrack. It utilizes a 350P op amp and delivers an output signal at a level of 0 dBm. (*Courtesy of Opamp Labs Inc.*)

STEREO PHONOGRAPH AMPLIFIER

A single LM1877 IC is used in the phonograph amplifier whose circuit is given in Figure 3-23. It will deliver up to 2 watts per channel. A ceramic stereo cartridge is used as the input signal source. Volume is controlled by

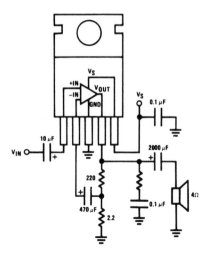

Figure 3-21. Single IC 8-Watt Audio Power Amplifier

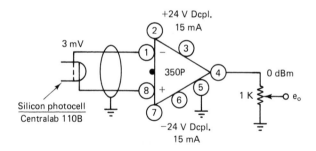

Figure 3-22. Sound-On-Film Preamplifier

the dual potentiometer at the inputs of the IC halves. A 100,000-ohm potentiometer connected between the output and the inverting input of each IC
half serves as a treble/bass booster/attenuator. (*Copyright National Semiconductor Corporation.*)

STEREO PHONO PREAMPLIFIER

A single LM1303 IC is used in the magnetic phonograph preamplifier
circuit illustrated in Figure 3-24. RIAA equalization is provided by the R-C
network connected to the output (pin 1) and inverting input (pin 6). Output
voltage swing is up to 5 volts RMS at 1000 Hz, with THD (total harmonic
distortion) of 0.1 percent. Voltage gain is 34 dB at 1000 Hz, and input overload
point is 100 millivolts RMS, also at 1000 Hz. The diagram shows only one
stereo channel; the second channel uses identical circuitry. Second channel

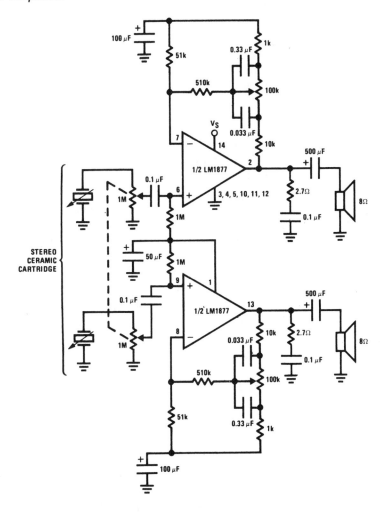

Figure 3-23. Stereo Phonograph Amplifier

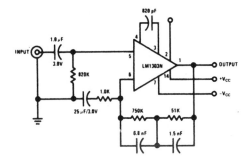

Figure 3-24. Stereo Phono Preamplifier

connections are made to pin 13 for output, pin 8 for inverting input, pin 9 for noninverting input (signal input), pins 11 and 12 for the 820-pF lag capacitor. The +15 volt DC supply is connected to pin 14 and the −15 volt DC supply is connected to pin 7 to serve both channels. (*Copyright National Semiconductor Corporation.*)

STEREO TAPE PREAMPLIFIER

The CA-3052 quadruple-amplifier IC can be used in a tape playback system. Each channel of this IC can be divided into two amplifier segments with external volume control and tone control. Two of the amplifiers can be connected in cascade as one channel of a stereo preamp, as shown in Figure 3-25. (All four amplifiers are used for two-channel stereo.) Each of the two-stage amplifiers has a gain of 46 dB (voltage gain of 200 times) and produces 1 volt of output. Total harmonic distortion (THD) is less than 0.17 percent. Separate bass and treble tone controls are connected between the two amplifier stages. Amplifier channel gain is controlled by a feedback-type volume control circuit in the second stage. Amplifier gain rather than input level is varied. The second channel of the preamplifier uses the two remaining stages of the IC and is connected in the same manner as the first. (*Courtesy of RCA Solid State.*)

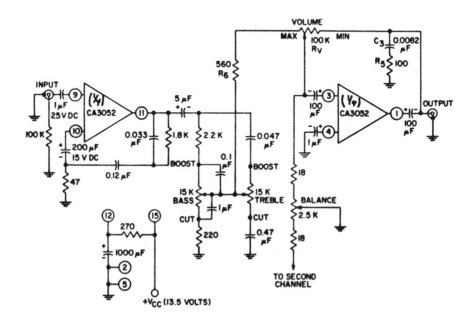

Figure 3-25. Stereo Tape Preamplifier

TRANSFORMER-COUPLED AMPLIFIER

Transformer-coupling of amplifier stages has the advantage of being able to match widely varying impedances. Figure 3-26 illustrates a two-stage transformer-coupled amplifier. The output from the collector of NPN transistor Q1 is fed directly to the primary winding of transformer T1. From the secondary of T1, the signal is fed to the base of Q2, the second common-emitter stage.

Bias for Q2 is provided by a divider network consisting of resistors R2 and R3. Capacitors C1, C2 and C3 provide AC signal coupling, while degenerative feedback is applied through unbypassed resistor R4.

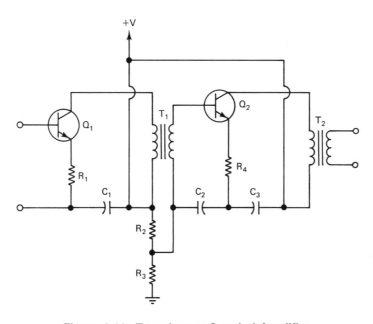

Figure 3-26. Transformer-Coupled Amplifier

100-WATT MONO AMPLIFIER

The circuit of a monaural AF amplifier capable of 100 watts output is given in Figure 3-27. It utilizes two power op amps hooked up in a bridge circuit. Note that the output is unbalanced with respect to ground. (Disconnecting the jumper at X permits a reduced output at the top terminal.) The speaker feed line should not be shielded. (*Courtesy of Opamp Labs Inc.*)

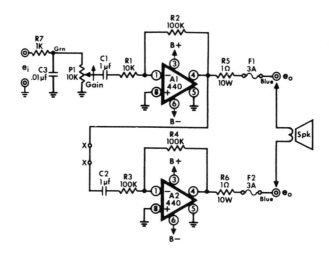

Figure 3-27. 100-Watt Mono Amplifier

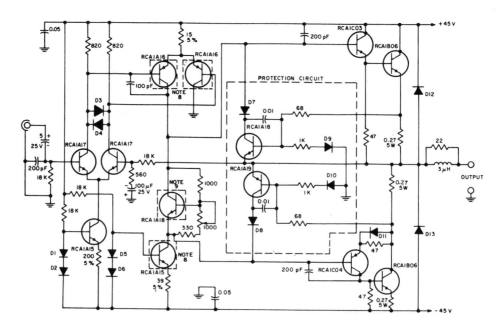

Figure 3-28. 70-Watt Quasi-Complementary-Symmetry Audio Amplifier

70-WATT QUASI-COMPLEMENTARY SYMMETRY AUDIO AMPLIFIER

A quasi-complementary-symmetry output circuit is used in the 70-watt audio amplifier circuit given in Figure 3-28. To provide push-pull output, an NPN driver transistor (1C03) is used to drive an NPN 1B06 transistor. And a PNP driver (1C04) is used to drive an NPN output transistor (1B06). With a 600-millivolt input signal fed to the base of one of the transistors (1A17) used in the differential input stage, this amplifier should deliver 70 watts into an 8-ohm speaker system, 100 watts into a 4-ohm load, or 50 watts into a 16-ohm load. Another differential amplifier (using a pair of PNP transistors) is used as a predriver; output from the collector of one of the transistors is direct-coupled to the NPN driver transistor. The PNP driver is fed from the collector of the NPN 1A15 transistor that is used as a current source circuit. A 1A18 and a 1A19 transistor plus five diodes (D7, D8, D9, D10, and D11) are used in an overload protection circuit. The speaker is fed through a 3-microhenry inductor shunted by a 22-ohm resistor from the junction of the emitters of the output transistors. Each of these emitters has an 0.27-ohm resistor in series to equalize current flow. (*Courtesy of RCA Solid State.*)

CHAPTER 4

Radio Frequency Amplifiers

INTRODUCTION

The radio frequency, or RF, amplifier uses many of the same principles found in audio amplifiers. Coupling between stages may be direct, through a combination of resistors and capacitors, or by means of transformers. Feedback is used to provide automatic gain control and to improve stability. And it is often necessary to use multiple stages of amplification to boost minuscule signals up to levels of hundreds of watts or more.

Still, there are additional considerations at radio frequencies. First, there is the matter of the bandwidth of the signals to be amplified. In this respect, there are two basic types of RF amplifiers: narrow-band and wide-band.

Narrow-band RF amplifiers are found in radio IF stages, RF power amplifiers, and in other applications where the bandwidth (the frequency range of signals to be amplified) is comparatively narrow. Each stage of a narrow-band RF amplifier often contains some kind of tuned circuit at both input and output.

In a narrow-band RF amplifier, the designer tries to maintain high Q in these input and output resonant circuits. (Q is a figure of merit—it represents the efficiency of a resonant circuit; that is, the amount of energy returned to the circuit $Q = \dfrac{X_2}{R}$.)

A wide-band RF amplifier is sometimes called a *video* amplifier. Because it must amplify signals over a wide frequency range, a wide-band RF amplifier often has some components present in *compensation* networks, to boost signals at the low or high frequency portions of its bandwidth. In a wide-band amplifier, the designer is trying to maintain good gain and flat frequency response, often over a bandwidth of several MHz or more.

While the output from an RF amplifier is the same frequency as the input, the fidelity of reproduction is seldom as important as it is with amplifiers at audio frequencies. For this reason, many RF amplifiers are operated class B push-pull for greater efficiency and power. RF amplifiers are even biased for class C operation, where high efficiency results from a transistor's or FET's being turned on for less than 50 percent of each cycle of input.

At radio frequencies, particularly in VHF and UHF amplifiers, the stray capacitances present in a circuit and its components can seriously affect its operation. In some cases, unwanted capacitance and parasitic oscillations must be suppressed by shielding or other techniques.

The RF characteristics of some semiconductors make them particularly useful in radio frequency amplifiers. Devices such as tunnel diodes, JFETs and hot carrier diodes are often used because of fast switching time, low internal capacitance, or high input impedance. In addition, a semiconductor RF amplifier may use configurations that actually reinforce its characteristics at high frequencies. For example, because of the type of internal capacitance, a transistor connected as a common-base amplifier has positive (regenerative) internal feedback at high frequencies, while in a common-emitter amplifier stage the feedback is degenerative. Inexpensive RF amplifiers sometimes take advantage of this by using a common-base configuration for extra gain. Also, a common-base stage has a larger natural bandwidth than the same transistor connected as a common-emitter amplifier.

This chapter contains a number of examples and practical circuits for semiconductor RF amplifiers.

AMPLIFIER WITH BASE FEEDBACK

The bandwidth of an amplifier can be broadened by using negative feedback from one stage to a preceding one. Figure 4-1 shows an amplifier consisting of two RC-coupled common-emitter stages. A feedback network consisting of capacitor C4 and resistor R5 couples back a signal from the emitter of Q2 to the input at the base of Q1.

With negative feedback, gain is sacrificed for increased frequency response. To help preserve some of the gain, resistor R9 is kept very low (about 10-25 ohms), while R5 is much higher (on the order of 1000-2000 ohms).

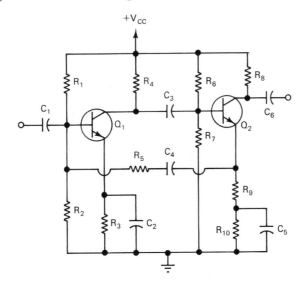

Figure 4-1. Amplifier with Base Feedback

AMPLIFIER WITH EMITTER FEEDBACK

Negative feedback from the output of the second stage to the emitter of the first is used to increase frequency response and stability of the amplifier shown in Figure 4-2. The amount of feedback is determined by the values of R6 and R5; decreasing the value of R6 increases the amount of feedback and decreases the gain of the amplifier.

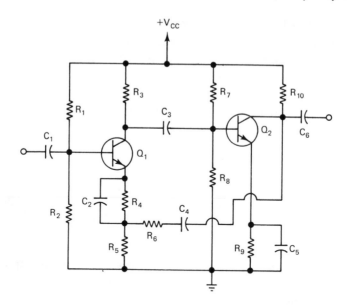

Figure 4-2. Amplifier with Emitter Feedback

AUTOMATICALLY CONTROLLED RF BOOSTER AMPLIFIER

The circuit of a 20-watt RF booster amplifier is given in Figure 4-3. It is designed to be driven by a 5-watt FM transmitter. The input signal is fed into it through the contacts of relay K2. When the relay is de-energized, the input signal is routed around the amplifier and is fed directly to the antenna through another set of contacts of relay K2 and a low pass filter (FL1) and J1. When the relay is pulled in the input signal is fed to the base of transistor Q1. Output from Q1 drives Q2 and Q3 which operate in parallel as the RF power amplifier. A magnetic reed relay (K1) has its coil connected in series with the 12 VDC power fed to the FM exciter. When the FM exciter is keyed on, the magnetic reed relay is pulled in and its contacts close and allow the 12 VDC power source to energize the RF booster amplifier. Diode D2 prevents operation of K2 if the power source polarity is reversed and diode D1 eliminates the inductive surge across the coil of K2. (*Courtesy of Aerotron, Inc.*)

BROADBAND COMMON-GATE FET RF AMPLIFIER

A CP640, CP664, CP665 or CP666 silicon epitaxial junction N-channel field effect transistor is used in a common-gate configuration in the broad band RF amplifier circuit given in Figure 4-4. These FETs can be used at frequencies up to 300 MHz and have a typical transconductance of 100,000 microsiemens (micromhos). In this circuit, source resistor R1 is selected to set drain current at 40 milliamperes when supply delivers 20 volts DC. (*Courtesy of Teledyne Semiconductor.*)

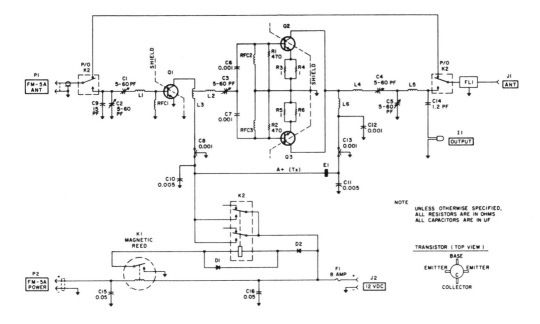

Figure 4-3. Automatically Controlled RF Booster Amplifier

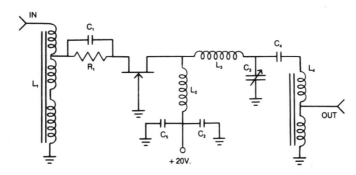

Figure 4-4. Broadband Common-Gate FET RF Amplifier

BROADBAND FET RF AMPLIFIER WITH CONSTANT CURRENT SOURCE

A CP640 or CP643 silicon epitaxial N-channel field effect transistor is used in common-gate configuration in the circuit given in Figure 4-5. In lieu of a source biasing resistor, a bipolar transistor (2N2222 or similar) provides a constant current source. A diode in the base voltage divider circuit conpensates for temperature variations that might affect the base-emitter voltage, which is set at approximately 1 volt. The input impedance is approximately 20 ohms when using the CP640 FET and 40 ohms when using the CP643 FET. (*Courtesy of Teledyne Semiconductor.*)

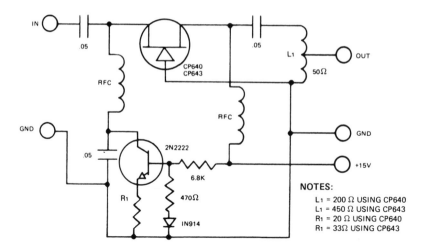

Figure 4-5. Broadband FET RF Amplifier with Constant Current Source

BROADBAND RF POWER AMPLIFIER FOR
FEEDING CARRIER CURRENT ANTENNA

The circuit shown in Figure 4-6 is of the transmitter section of a wireless intercom system. This circuit shows how the output of a radio transmitter may be connected to an AC power line used as the transmission medium. The system is designed for operation in the 100 or 200 kHz frequency band. One problem with this system is that in this frequency range, transmission losses from the power line can be excessive compared to those along a dedicated transmission line.

In the diagram, T1 is the RF amplifier tank which is broadly tuned for maximum ouput power by varying the inductance of its core. C8, which is in series with the antenna coil, has a capacity of 0.1 μF. Because of the high value of C8, it is not practical to use a variable capacitor for loading of the final RF stage to the wire antenna line. However, selection of values for C8 will give optimum performance.

The frequency is generated by a free-running VCO in the LM566CN timer IC (IC1). (Line coupling coils for use as L1 are available from Toko America Inc.) Values of C4 and C7 for 100 kHz or 200 kHz operation are shown on the drawing. Resistors R7 and R8 may be selected to obtain a frequency response out to 20 kHz. This will make the circuit suitable for music transmission. The receiver can be a TRF type using the same coils and shunt capacitors as those used for T1, C4, and C7 in the transmitter. Or an up-converter can be used to translate the transmitter frequency to above 30 MHz, making it possible to utilize a scanner receiver or a low-band land mobile receiver.

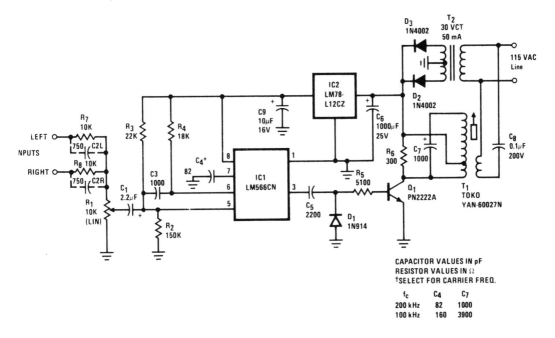

Figure 4-6. Broadband RF Power Amplifier for Feeding Carrier Current Antenna

This particular transmitter generates a narrow-band FM signal. At the frequency of operation, this signal is significantly less noisy than an AM signal would be, and by controlling modulation level, band occupancy can also be controlled. This transmitter may be operated without an FCC license if it is certified to comply with the applicable technical standards of Part 15, FCC Rules and Regulations.

For full duplex communication, the west end transmitter could be at 100 kHz and the east end could be at 200 kHz. The receiver at each end of the line would then be operated at the transmission frequency of the opposite end.

Transmission distance can vary greatly, from 100 feet to more than a mile. When there is a distribution transformer between the power line ends, the system should avoid having any unit on the opposite leg of a transformer, because this may introduce excessive attenuation of RF signals. (*Courtesy of Radio Shack, a division of Tandy Corp.*)

BROADBAND VHF VMOS RF POWER AMPLIFIER

An RF power amplifier circuit using a VMOS transistor is given in Figure 4-7. This amplifier covers a frequency range from roughly 60 MHz to 260 MHz with a minimum gain of 10 dB. When a Siliconix VMP-4 VMOS transistor is used, the spot noise figure should be lower than 2 dB at 150 MHz. Transformer T1 may consist of four twisted pair turns wound on an

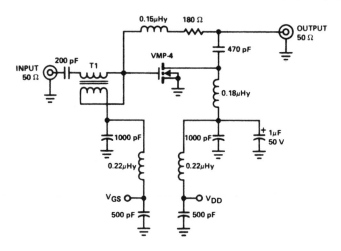

Figure 4-7. Broadband VHF VMOS RF Power Amplifier

Indian General F625-902 coil form. The drain supply voltage (V_{DD}) may be up to 24 volts DC and the gate bias (V_{GS}) may be set at up to +5 volts DC. Drain current will be between 400 and 600 milliamperes when gate bias is +5 volts DC and drops off to 10 microamperes when gate voltage is zero.

This circuit can be used as a receiver preamplifier or as an RF power or driver amplifier in a transmitter. It is capable of delivering an RF output of 1 watt into 50 ohms. (© *Siliconix incorporated.*)

CASCODE VHF RECEIVER PREAMPLIFIER

The circuit given in Figure 4-8 is a preamplifier for an external VHF receiver. This cascode, field-effect transistor amplifier provides 18-20 dB of gain with a low noise figure. The slugs in the coils should be peaked to compensate for reactances in the external circuit. The cascode preamp should not oscillate under any tuning conditions.

CLASS C RF AMPLIFIER STAGE

In a transistor amplifier which is biased for class C operation, collector current flows for less than half a cycle of input. The output, then, is not even close to being a faithful replica of the input signal, and class C amplifiers are not useful for audio reproduction. However, because a transistor (or FET) is on for only a fraction of time, class C amplifier stages are highly efficient, and they are often used to amplify radio frequencies by means of "ringing."

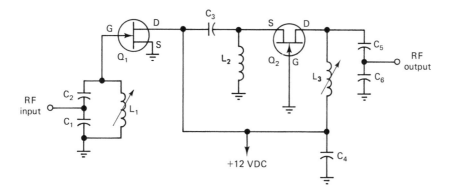

Figure 4-8. Cascode VHF Receiver Preamplifier

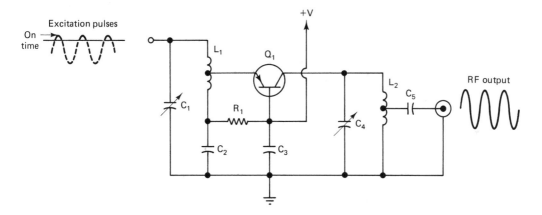

Figure 4-9. Class C RF Amplifier Stage

Figure 4-9 shows a simple, common-base RF amplifier stage. Transistor Q1 is biased so that it is turned on for only a small portion of each cycle of input. The *conduction angle* (the portion of input cycle during which collector current flows) is about 90-120° for most RF amplifiers using class C biasing.

The output tank circuit, consisting of capacitor C4 and coil L2, is tuned to the same frequency as the input signal. The brief pulses of collector current cause this tank circuit to oscillate, or "ring," at the same frequency as the input, producing an amplified RF signal at the output. (In some circuits, depending on their design, a "ringing" output may not be the same frequency and phase as the input.)

Because the driving element is on for only a small portion of each cycle, this type of circuit can achieve 85 percent efficiency.

COLOR VIDEO AMPLIFIER

So that a color video signal is accurately reproduced by an amplifier, the phase error through the amplifier stage must be kept small. Figure 4-10 shows a video amplifier whose differential phase error is less than 2°. The circuit uses a Signetics NE 5539 chip, with the input signal fed into the noninverting input of the IC. Frequency compensation (pin 12) and offset adjustment (at the inverting input) are optional. With a V of ± 8 volts and R_F of 1.5 kilohms, gain of the amplifier is 28 dB. (*Courtesy of Signetics Corporation.*)

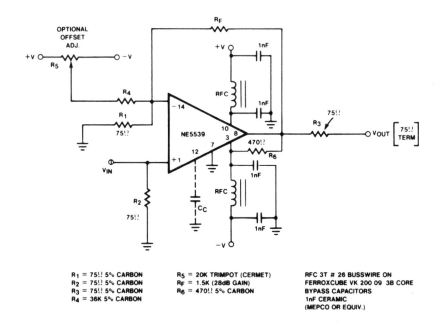

R_1 = 75Ω 5% CARBON	R_5 = 20K TRIMPOT (CERMET)	RFC 3T # 26 BUSSWIRE ON
R_2 = 75Ω 5% CARBON	R_F = 1.5K (28dB GAIN)	FERROXCUBE VK 200 09 3B CORE
R_3 = 75Ω 5% CARBON	R_6 = 470Ω 5% CARBON	BYPASS CAPACITORS
R_4 = 36K 5% CARBON		1nF CERAMIC
		(MEPCO OR EQUIV.)

Figure 4-10. Color Video Amplifier

FM RECEIVER IF STRIP

The circuit of a 95-dB gain, 10.7-MHz IF amplifier strip for use in FM broadcast band receivers is given in Figure 4-11. The mixer output is fed through IF transformer T1 to the base of the input stage NPN transistor, whose output in turn is coupled through T2(a) to the second stage NPN transistor. The output of this stage is coupled through T2(b) to the third stage NPN transistor. And the output of the third stage is fed through T3 to the ratio detector, which utilizes a matched pair of 1N542 diodes. The audio output to an FM/MPX stereo demodulator is taken from across C10. (*Courtesy of RCA Solid State.*)

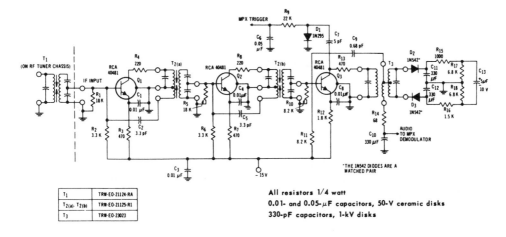

Figure 4-11. FM Receiver IF Strip

T_1	TRW-EO-21124-RA
$T_{2(a)}$, $T_{2(b)}$	TRW-EO-21125-R1
T_3	TRW-EO-23023

All resistors 1/4 watt

0.01- and 0.05-μF capacitors, 50-V ceramic disks

330-pF capacitors, 1-kV disks

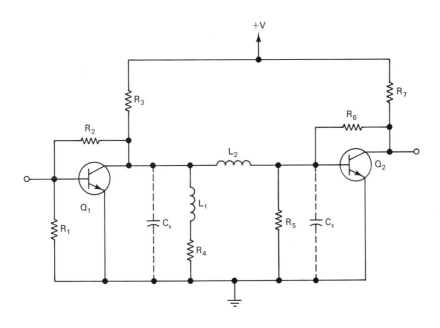

Figure 4-12. High-Frequency Peaking Coils

HIGH-FREQUENCY PEAKING COILS

The bandwidth of an amplifier stage can be broadened by the use of high-frequency peaking coils. As shown in Figure 4-12, coil L1 has been placed in series with load resistor R4. This effectively raises the load impedance at high frequencies, causing increased HF response. Because L1 is not in series with the signal path, it is known as a shunt peaking coil.

Series peaking coil L2 has been added in series with the signal line to tune the stray capacitances, represented on the diagram by C_s, at the output of Q1 and the input of Q2. This coil and the stray capacitances form a pi-network which further increases the bandwidth of the amplifier.

HIGH POWER RF AMPLIFIER

The circuit given in Figure 4-13 is of the RF power amplifier section of the Hull 255 SSB transmitter. This amplifier can operate at any frequency in the 2-24 MHz range without retuning. As used in the Hull 255 SSB marine radio transceiver, it delivers 150 watts PEP output at frequencies below 3 MHz and 225 watts PEP at frequencies within the 3-24 MHz range.

The output of the transmitter driver (not shown) is fed to the base of Q701 through R701 and C701. This single-ended predriver stage feeds its output to the push-pull driver stage which utilizes Q703 and Q702. The push-pull output of the driver stage is fed through T702 to the bases of Q704 and Q705. The output of the final RF amplifier stage is inductively coupled through T704 to the antenna system through a harmonic filter and antenna tuner (not shown) which automatically tunes the antenna to the selected operating frequency.

The power source is 13.8 volts DC which is fed directly to the collectors of the transistors in the driver and output stages. The collector voltage to Q701 passes through R705. The forward bias for Q701 is obtained from the junction of R704 and R705. The forward bias of Q702 and Q703 is obtained from the 13.8-volt bus through R710 and the chokes L701 and L702. The bias load is limited by diode CR701. The bias for Q704 and Q705, the output transistors, is fed through the base-emitter path of Q706 whose base-collector voltage is limited by diodes CR702 and CR703 which are connected in series between the base of Q706 and the ground bus. The output stage bias can be varied by adjusting potentiometer R716. Ferrite beads are installed on the leads of feedback resistors R709 and R708, as well as on the bias lead for the output stage. These ferrite beads suppress parasitic oscillations. Generous amounts of feedback are provided in all three stages.

All of the RF transformers (T701, T702, T703 and T704) are broadband types. The three-winding transformers isolate the output from the power supply circuit and balance the output tank T704 with respect to ground. (*Courtesy of Hull Electronics Company.*)

JFET CATV AMPLIFIER

The circuit of a three-stage wideband amplifier is given in Figure 4-14. This amplifier uses three U310 JFETs in the common gate configuration. Gain is around 15 dB per stage at 200 MHz. The component values listed in

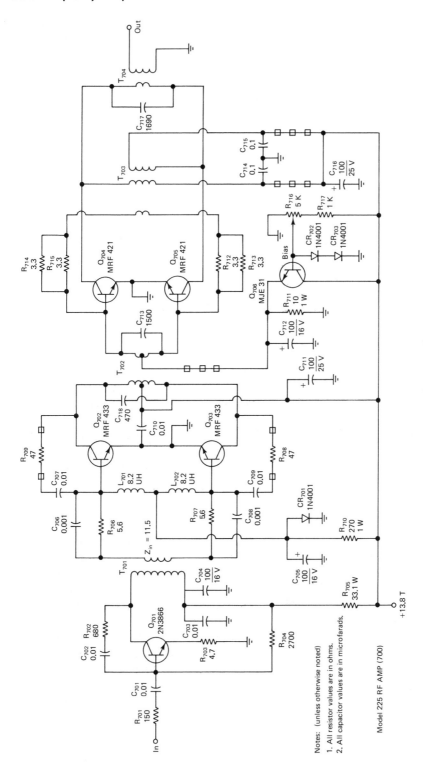

Figure 4-13. High Power RF Amplifier

the diagram are for the 200-250 MHz range. By changing the values of the inductors, the frequency range can be lowered or raised. (© *Siliconix incorporated.*)

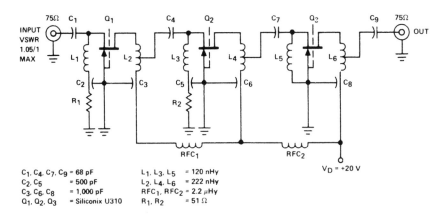

$C_1, C_4, C_7, C_9 = 68$ pF
$C_2, C_5 = 500$ pF
$C_3, C_6, C_8 = 1,000$ pF
$Q_1, Q_2, Q_3 = $ Siliconix U310

$L_1, L_3, L_5 = 120$ nHy
$L_2, L_4, L_6 = 222$ nHy
$RFC_1, RFC_2 = 2.2$ μHy
$R_1, R_2 = 51\ \Omega$

Figure 4-14. JFET CATV Amplifier

900-MHz RF POWER AMPLIFIER

The circuit of a 10-watt RF power amplifier for use in the 900-MHz UHF region is given in Figure 4-15. It utilizes an MRF824 transistor to which the input signal is fed through Z1, a 50-ohm microstrip line 0.45 centimeters wide. The output is fed through Z2 which is identical to Z1. This amplifier will deliver 10 watts of output power into a 50-ohm load when its input driving power is between 2 and 3 watts and when operated from a 12.5-volt DC supply. This type of amplifier is ideal for personal radio service transmitters in the 900-MHz region. (*Courtesy of Motorola Inc.*)

NEUTRALIZED AMPLIFIER STAGE

At radio frequencies, the small capacitances and resistances present in a transistor, FET, or other semiconductor device begin to loom large. Especially in RF power amplifier stages, the feedback coupled through stray capacitance can be enough to cause a stage to go into oscillation. The solution is to add *neutralization* through a capacitor (or, at extremely high frequencies, an inductor).

Figure 4-16 shows an NPN transistor connected in a common-emitter RF amplifier stage. Input to the stage is fed through transformer T1 to the base of transistor Q1, and output is taken off the collector and coupled through T2 to the next stage. The dots on T2 show the ends of the transformer windings whose signals are in phase; thus the signal coupled back to the base through neutralizing capacitor C2 is negative feedback, 180° out of phase with the input.

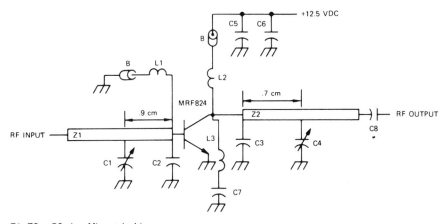

Z1, Z2   50-ohm Microstrip Line
          .45 cm Wide Board ϵ_R 2.56
          Board Thickness .157 cm

Capacitors — Spacing to Centers   Inductors

C1, 4   Johanson 1–10-pF, No. 5201   L1   6T, No. 24 AWG .25 cm ID X .6 cm Long
C2, 3   ATC 10-pF, 50-Mil Chip   L2   5T, No. 22 AWG .3 cm ID X 1 cm Long
C5   Underwood 40 pF   L3   3T, No. 28 AWG .25 cm ID X .6 cm Long
C6   1 μF, 35 V Tantalum   B   Ferroxcube 4B Ferrite Bead
C7   Erie .1 μF 100 V
C8   ATC 39-pF 100-Mil Chip

Figure 4-15. 900 MHz RF Power Amplifier

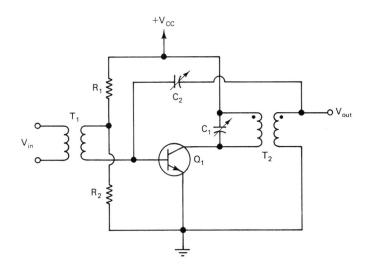

Figure 4-16. Neutralized Amplifier Stage

To adjust the amount of neutralization, disconnect the primary of T2 from V_{CC}, then adjust C1 for maximum output and C2 for minimum output at V_{OUT}. For best results, these two adjustments should be "rocked" several times.

It is important to remember that the resistance of a resistor should not change with frequency and that the reactance of inductors and capacitors does. At UHF, the resistor leads exhibit inductance and capacitance. Therefore, all resistor leads should be as short as possible.

PUSH-PULL PARALLEL VMOS VHF RF POWER AMPLIFIER

Two VN66AJ VMOS transistors are used in a push-pull parallel configuration in the circuit given in Figure 4-17. This circuit is of a 2-meter band RF power amplifier. The gates of the VMOS devices are fed in parallel through independent matching networks. Their drains are fed to a push-pull output circuit, consisting of L3 and L4. The output is taken from one end of the push-pull output circuit through the 12 pF capacitor; DC operating power is fed through a 1 microhenry RF choke to the other end of the push-pull output circuit.

The amplifier can be balanced by adjusting the 500-ohm potentiometers which individually control the forward bias on the gates of the VMOS transistors. (© *Siliconix incorporated.*)

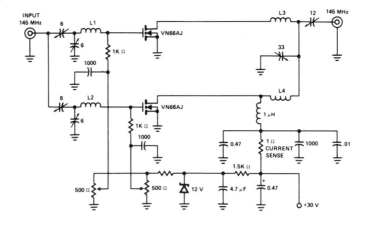

Figure 4-17. Push-Pull Parallel VMOS VHF RF Power Amplifier

UHF POWER AMPLIFIER

The RF power output section of most solid-state VHF and UHF transmitters usually consists of two or three cascaded stages employing bipolar power transistors. An example of a typical circuit is given in Figure 4-18. The input signal from the exciter is fed in through C1 and L1 to the base

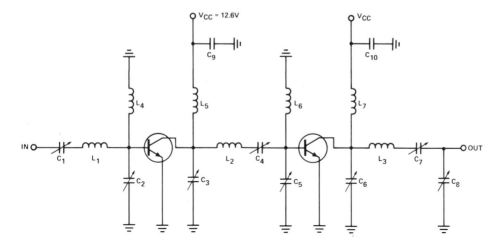

Figure 4-18. UHF Power Amplifier

of the first transistor. The output signal of the first transistor is fed through L2 and C4 to the base of the output stage transistor, whose output in turn is fed through L3 and C7 to the antenna system through a low-pass filter (not shown). For operation in the 400-MHz range, variable capacitors C1 through C8 may be ceramic trimmers with a maximum capacitance of 10 pF. Coils L1, L2 and L3 consist of a half-turn of tin-plated copper wire, 20 millimeters long and 0.6 millimeter thick. Coil L4 consists of six turns of 0.2 millimeter thick enamel-coated copper wire wound on a 33-ohm resistor. L5 and L7 each consist of two turns of 0.6 millimeter thick tin-plated copper wire with the inside diameter of the coil being 6 millimeters. L6 consists of six turns of 0.2 millimeter thick enamel-coated copper wire wound on a 10-ohm resistor. The series-resonant networks at the input and output and between the two transistors enable matching the relatively low impedances to a 50-ohm source and to a 50-ohm load.

UHF RECEIVER PREAMPLIFIER

Many UHF communications receivers are supplied without an RF amplifier ahead of the mixer when the added gain is not required and to avoid potential modulation interference problems. Some manufacturers offer an optional RF preamplifier assembly which fits inside of the receiver. A schematic of the RF preamplifier used in Aerotron and Quintron UHF receivers is given in Figure 4-19. The input signal from the antenna (through a preselector) is fed through C2001 to the gate of a 2N5398 FET. The input of the FET is shunted by L2001 (½ turn, No. 20 copper wire of 0.55-inch diameter) and with C2002, an air dielectric capacitor, for maximum gain. L2003 is a variable inductor in the neutralization loop. The output signal is developed across L2002 which consists of two turns of No. 20 copper wire with an

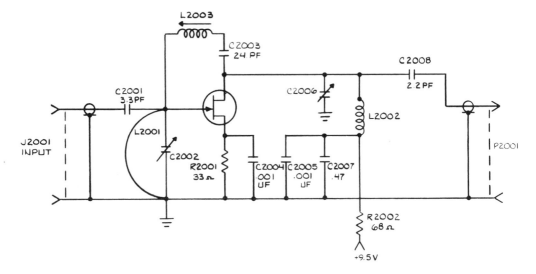

Figure 4-19. UHF Receiver Preamplifier

inside diameter of 0.156 inch. L2002 is tuned to resonance with C2006, an air dielectric 1.3-5.4 pF variable capacitor. (*Courtesy of Quintron Corporation.*)

UHF RF POWER AMPLIFIER

The circuit of a 35-watt RF power amplifier for the 406-512 MHz range is given in Figure 4-20. It utilizes a CM45-12A power transistor and is fed by a 10-watt exciter. The input circuit utilizes microstrips in lieu of conventional coils to serve as impedance matchers and as resonant circuits. Input tuning is accomplished by adjusting variable capacitor C2. Output tuning is accomplished by adjusting C10 and C11. A built-in sensor is used for sensing reflected power, which is rectified by hot-carrier diode CR1. Forward power is also sensed and is rectified by diode CR2. The DC outputs of CR1 and CR2 are fed to an external metering circuit. This circuit enables determination of antenna system SWR by measuring the forward and reflected power. The RF amplifier is powered by a 13.6 VDC source and its RF output is fed through a harmonic filter (not shown) to a 50-ohm antenna system. (*Courtesy of Quintron Corporation.*)

VERTICAL DEFLECTION AMPLIFIER FOR OSCILLOSCOPE

The circuit of a solid-state vertical deflection oscilloscope amplifier with a 3 dB bandwidth of DC to 375 MHz is given in Figure 4-21. Its gain is rated at 46 dB and vertical sensitivity is 10 mV/cm. The input signal is fed to the base of TR2 through R4. This amplifier utilizes direct coupling between all

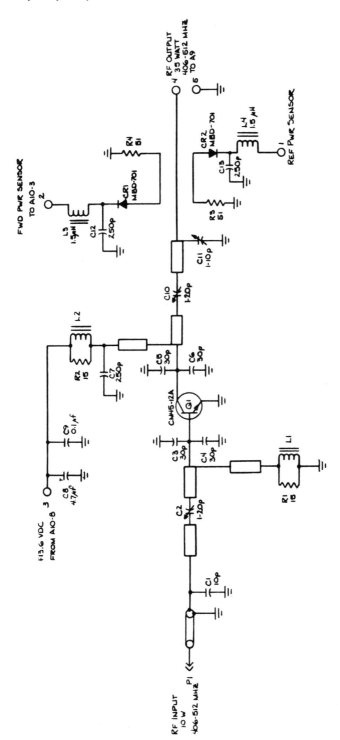

Figure 4-20. UHF RF Power Amplifier

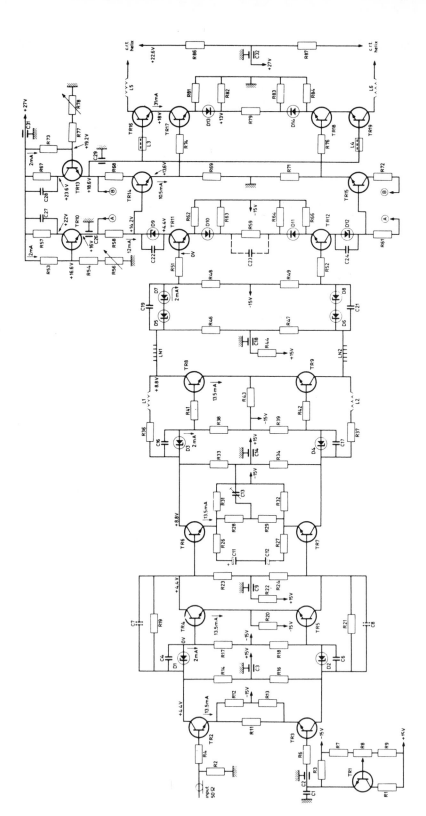

Figure 4-21. Vertical Deflection Amplifier for Oscilloscope

stages, and Zener diodes are used to stabilize the bias on the transistors. The first stage functions as a phase inverter. The next five stages operate in complementary symmetry push-pull. Ferrite beads (L3-L6) on the leads to the base of TR16 and TR19 eliminate parasitic oscillations. The collector of TR16 is connected to one of the CRT's vertical deflection coils (L5) and TR19 to coil L6. This amplifier is designed to utilize A430 and A485 transistors. (*Courtesy of Amperex Electronics Corp.*)

VMOS VHF RF POWER AMPLIFIER

A single VN66AJ VMOS transistor is capable of delivering nearly 5 watts PEP output at 146 MHz when used in the circuit given in Figure 4-22. The power supply delivers +30 volts DC to the drain of the power FET. This voltage is reduced by the resistor in series with the Zener diode, and by the voltage divider that follows the Zener regulator diode to provide a low DC voltage for forward-biasing the gate of the VMOS transistor. The input can be tuned with the series and shunt variable capacitors to match the output impedance of the driver, and the output impedance can be adjusted with the series and shunt capacitors ahead of the antenna jack to match into a 50-ohm antenna system. (© *Siliconix incorporated.*)

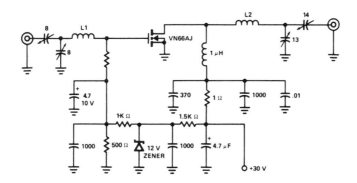

Figure 4-22. VMOS VHF RF Power Amplifier

VOLTAGE-CONTROLLED VARIABLE GAIN AMPLIFIER

A trio of FETs are used as voltage-controlled resistors in the voltage-controlled variable gain amplifier shown in Figure 4-23. The series drain-source resistance of the two FETs at the top of the diagram is varied with the potentiometer whose wiper is connected to the gates of these two FETs, and the shunt drain-source resistance of the third FET is varied with the other potentiometer. (© *Siliconix incorporated.*)

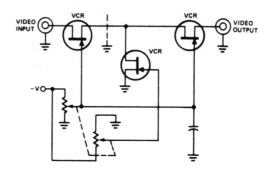

Figure 4-23. Voltage-Controlled Variable Gain Amplifier

CHAPTER 5

Oscillators and Frequency Generators

INTRODUCTION

When enough positive, or regenerative, feedback is applied to an amplifier, the circuit goes into *oscillation*—that is, it produces an output even when there is no input. The circuit then becomes an oscillator. By controlling the frequency of oscillation and the shape of the waveform produced, a designer can use the phenomenon of oscillation to generate signals for a variety of purposes.

The electronic oscillator was discovered accidentally in 1912 by Dr. Lee De Forest and his associates in Palo Alto, California, while they were developing an audio amplifier. When the output transformer was placed close to the input transformer, the inductive coupling between the two transformers provided a positive feedback path, and oscillations were generated. At about the same time, Major Edwin H. Armstrong deliberately designed an oscillator utilizing De Forest's audio (triode) tube.

Many oscillators, both tube and solid-state type, are designed to produce output that is a sine wave. Others produce nonsinusoidal waveforms such as square waves, trapezoidal patterns, spikes, etc. Nonsinusoidal oscillators include the multivibrator, the relaxation oscillator, and others.

Among nonsinusoidal oscillators, rate of oscillation ("frequency" in sinusoidal circuits) is sometimes referred to as pulse repetition frequency (PRF) or pulse repetition rate (PRR). This is nothing more than the number of pulses per second the oscillator produces. Another important parameter may be rise time. This is the length of time that it takes a leading edge of a pulse to rise from 10% to 90% of its amplitude.

Among LC and RC sinusoidal oscillators, there are those which produce oscillation by phase-shifting networks. In a phase-shift oscillator, the output of an amplifier (which is normally 180° out of phase with the input) is shifted another 180° and reapplied to the input. The effect is of a signal that is in phase with the input, resulting in oscillation. An example of a phase-shift oscillator is the Wien bridge configuration.

Another type of oscillator, also usually sinusoidal, results from the "ringing" of a tuned tank circuit. These LC oscillators include the Clapp, Colpitts, Hartley, and Reinartz configurations. In this type of oscillator, LC components determine the frequency of oscillation, while a driving element, such as a transistor or FET, prevents oscillations produced in the tank from being damped out.

The function of an oscillator is seldom to deliver a strong signal—other circuitry is present for amplification—but the oscillator configuration is sometimes chosen because of its output level. A more important consideration is an oscillator's stability: how much the frequency of oscillation varies with temperature or load. Generally, an oscillator whose driving element is tightly coupled to its frequency-determining network will give better signal strength but poorer stability than one whose driving element and frequency-determining network are loosely coupled.

The addition of a piezoelectric crystal to an oscillator can markedly improve the circuit's stability. The crystal may be wired in series or in parallel. In operation, the crystal behaves very much like a stable, high-Q parallel resonant circuit.

Virtually all oscillators may be used as frequency generators, although in a frequency generator stability is usually of paramount importance.

ARMSTRONG OSCILLATOR

Positive feedback in the Armstrong oscillator is obtained through a tickler coil near the main tank coil. In Figure 5-1, the frequency of oscillation is determined by the L1C1 tank. Feedback to sustain oscillation is coupled from the collector of transistor Q1 into the tank through tickler winding L2. Capacitor C1 and resistors R1 and R2 form a bias network that prevents overdriving of Q1.

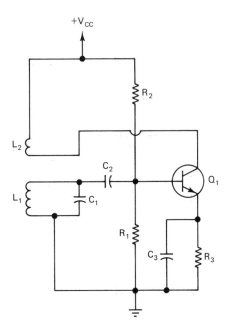

Figure 5-1. Armstrong Oscillator

ASTABLE MULTIVIBRATOR

The astable multivibrator is a nonsinusoidal oscillator that can produce a square-wave output. An astable multivibrator uses a pair of driving elements, with the output from one element providing feedback to the other.

The driving elements in the astable multivibrator of Figure 5-2 are NPN transistors Q1 and Q2. Each transistor has its collector coupled to the base of the other through a capacitor. Diodes D1 and D2 are not necessary for oscillation, but improve the rise time of the square waves that are generated by the multivibrator.

Assume that Q1 is turned on and is saturated. Q2 is cut off, and its collector is near the $+V_{cc}$ potential, while the collector of Q1 is near ground. C1 has been charged on the previous half-cycle.

With Q1 turned on, C2 charges through the base circuit of Q1 and resistor R5; at the same time, C1 discharges through R3. When C2 has been charged and C1 discharged, Q1 is cut off by the potential on C2, and Q2 is allowed to turn on, reversing the process. C1 now charges through R2 and the base circuit of Q2.

Because of diodes D1 and D2, the discharge paths for capacitors C1 and C2 are not through collector resistors R1 and R6, so each transistor can switch almost instantaneously. This produces a sharp square wave at the output. Frequency is primarily determined by the values for RC circuits C2R5 and C1R2.

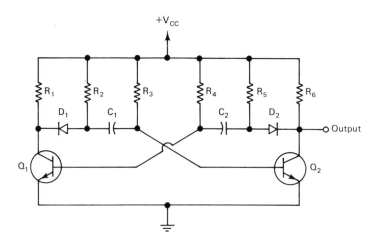

Figure 5-2. Astable Multivibrator

ASTABLE MULTIVIBRATOR USING A SCHMITT TRIGGER IC

An astable (free running) multivibrator utilizing the Sprague ULN-3303M IC is connected as illustrated in Figure 5-3. This IC contains two Schmitt triggers. Selecting a resistor and a capacitor allows setting the frequency between 4 Hz and 100 kHz. The output signal is a square wave. (*Courtesy of Sprague Electric Company.*)

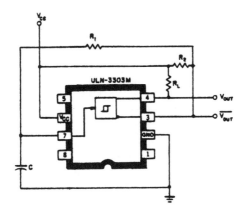

Figure 5-3. Astable Multivibrator Using a Schmitt Trigger IC

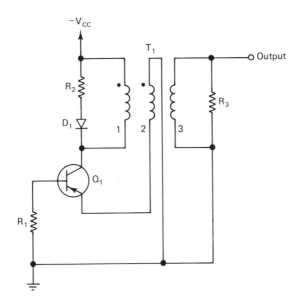

Figure 5-4. Blocking Oscillator

BLOCKING OSCILLATOR

Figure 5-4 shows a PNP transistor connected with a three-winding transformer to form a blocking oscillator. This type of circuit can be operated as a free-running, nonsinusoidal oscillator or as a triggered square-wave generator.

When Q1 is turned on, feedback from winding 2 of transformer T1 to the emitter holds the transistor on, and Q1 goes into saturation. At saturation, current through winding 1 is no longer changing, with the result that no voltage is induced in winding 2. This, in turn, allows the transistor to come quickly out of saturation, and the counter-emf induced by the collapsing lines of flux in the transformer now turns off Q1. When all lines of flux have collapsed, the V_{CC} potential again turns the transistor on, and the process repeats.

Transformer T1 is often a saturable reactor with a very square hysteresis characteristic. In the case of a transformer with a square hysteresis loop, the on time of the transistor is determined primarily by the transformer.

CALIBRATION OSCILLATOR

An FET is used as the oscillator in the calibrator whose circuit is given in Figure 5-5. Its output signal is fed to MFC 6020 IC, whose output in turn is utilized as a signal source. The oscillator operates at 100 kHz since it employs a crystal for that frequency. This frequency is divided by 4 in the IC. This calibration oscillator can be used to check the accuracy of the tuning dial of a receiver at 25-kHz intervals, or as a testing instrument at all frequencies from 25 kHz to 30 MHz. (*Courtesy of Hallicrafters.*)

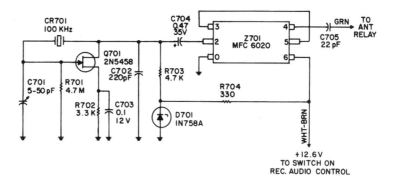

Figure 5-5. Calibration Oscillator

CLAPP OSCILLATOR

Like the Colpitts oscillator, the Clapp oscillator configuration has two capacitors in a tank circuit, with feedback for oscillation being taken from the junction of the capacitors. As shown in Figure 5-6, the identifying feature of the Clapp circuit is a third capacitor, C1, in series with the inductor and the twin capacitors C2 and C3. For this reason, the Clapp circuit is sometimes called a series-tuned Colpitts oscillator.

The circuit shown uses a JFET in a common-drain configuration, but a bipolar transistor or other driving element could easily be used. Frequency

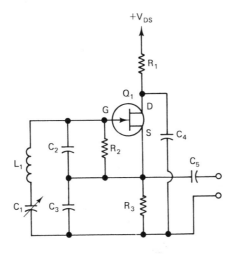

Figure 5-6. Clapp Oscillator

of oscillation is determined by all four series components in the tank, and biasing is provided by resistors R1, R2 and R3. Capacitor C4 is a short to ground at the frequency of oscillation.

Tank current in the Clapp circuit is lower and coupling looser than in the Colpitts oscillator, with the result that drift is lower and stability greater when the Clapp oscillator is used.

This circuit is also sometimes called a Gouriet oscillator.

COLPITTS OSCILLATOR

A basic Colpitts oscillator using an NPN transistor in a common-emitter configuration is shown in Figure 5-7. The salient feature of the Colpitts oscillator is a double capacitor (C3 and C4). Feedback for oscillation is picked off from the junction of the two capacitors. The circuit shown uses an inductor to form a tank circuit, but in some cases a crystal may be substituted for the inductor. Frequency of oscillation is determined by the values for the tank components: C3, C4 and L1. Bias is provided by all four resistors, and capacitor C1 is a virtual short to the desired AC signal.

COLPITTS OSCILLATOR UTILIZING A CA3048 IC

One of the four amplifiers contained in a CA3048 IC can be used in the Colpitts oscillator circuit given in Figure 5-8. The values noted on the diagram will cause the oscillator to operate at 33.546 kHz when the supply voltage is 9 volts and at 33.536 kHz when operated from a 12-volt DC supply. The ratio of C3 to C4 determines the amounts of signal fed back to the two inputs. C1 and C2 serve as DC blocking capacitors. (*Courtesy of RCA Solid State.*)

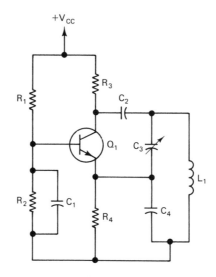

Figure 5-7. Colpitts Oscillator

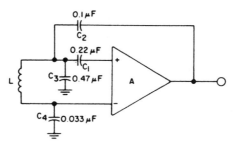

A IS ANY AMPLIFIER OF THE CA3048

Figure 5-8. Colpitts Oscillator Using a CA3048 IC

CONSTANT OUTPUT AUDIO OSCILLATOR

Four bipolar transistors and an FET are used in the 7-kHz oscillator circuit shown in Figure 5-9. This circuit is used in the Heathkit IM-5248 intermodulation distortion analyzer. Transistors Q101 and Q102 form a direct-coupled amplifier whose output is fed to the base of Q104 which is wired as an emitter-follower. The output signal from the emitter of Q104 is fed back to the base of Q101 and a bridge circuit consisting of R101, R102, R103, C101 and C102, which determines the oscillating frequency. The signal at the emitter of Q104 is fed through diode D101 and potentiometer R108 to the gate of Q103. The negative DC voltage across C104, which is determined by the signal level from Q104, varies the bias on the gate of Q103. The output

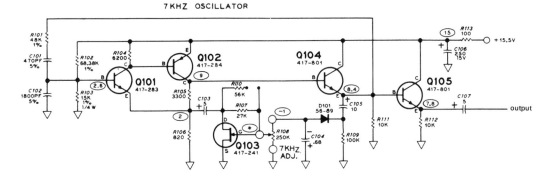

Figure 5-9. Constant Output Audio Oscillator

level may be varied with R108 and is stabilized by Q103. The output signal at the emitter of Q104 is also fed to the base of Q105 which serves as a buffer. The output impedance at the emitter of Q105 is 1000 ohms. (*Courtesy of Heath Company.*)

CRYSTAL-CONTROLLED DTMF ENCODER

The dual-tone multi-frequency encoder circuit given in Figure 5-10 is of the Heathkit Model HD-1984 Micoder II. It is designed for use with a radio transmitter or transceiver. The tones are generated by IC101, an MK 5086 IC which contains an oscillator whose frequency is determined and stabilized by a 3563.795-kHz quartz crystal. This oscillator is connected to IC pins 7 and 8, and is divided down to the required audio frequencies. The divider in IC101 divides the crystal frequency signal to the 687-1477 Hz range. The frequency division ratio is controlled with the pushbuttons on the keyboard. Each of the pushbuttons causes a different combination of tones to be generated. (*Courtesy of Heath Company.*)

CRYSTAL-CONTROLLED HARTLEY OSCILLATOR

Figure 5-11 shows a basic Hartley oscillator using a PNP transistor. The Hartley oscillator can be identified by its tapped inductor (L) in the tank circuit. Feedback is taken off the inductor tap and applied to the emitter of the transistor to maintain oscillation. Frequency is determined by the values for L and C. The piezoelectric crystal in the base circuit is also set to the oscillator frequency and helps make the circuit more stable.

DIRECT FM CRYSTAL-CONTROLLED OSCILLATOR

The circuit of a Colpitts oscillator with a varactor diode connected in series with the crystal to enable frequency modulating of the oscillator is

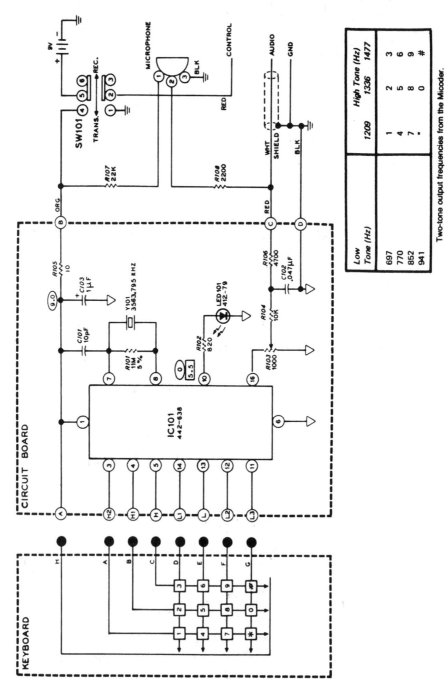

| Low | High Tone (Hz) | | |
Tone (Hz)	1209	1336	1477
697	1	2	3
770	4	5	6
852	7	8	9
941	*	0	#

Two-tone output frequencies from the Micoder.

Figure 5-10. Crystal-Controlled DTMF Encoder

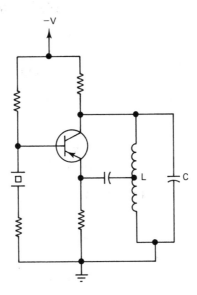

Figure 5-11. Crystal-Controlled Hartley Oscillator

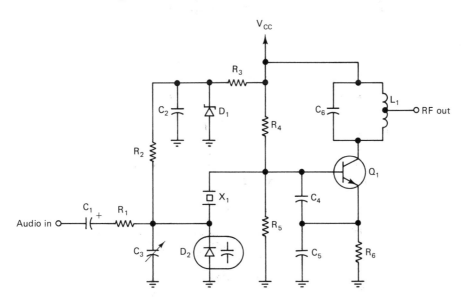

Figure 5-12. Direct FM Crystal-Controlled Oscillator

given in Figure 5-12. The varactor is biased to set the carrier frequency by a DC voltage obtained from the power source through R3 and R2. When the level of the audio modulating signal fed to the varactor through C1 and R1 is + 3 volts and the DC bias is − 7 volts, the net voltage excursions applied to the varactor will vary from − 4 volts (when AF level is + 3 volts) to − 10 volts (when AF level is − 3 volts) and the FM deviation will be around 1.7 kHz. When multiplied three times by a frequency tripler, an FM deviation of + 5.1 kHz is obtained at a frequency three times the oscillator frequency. This oscillator can be used in an FM transmitter or in an FM signal generator for checking FM receivers. It can be equipped with a crystal in the 10-17 MHz range. FM deviation will be greater at 17 MHz than at 10 MHz because the varactor capacity changes will be more significant. This circuit is based on the paper by Stuart J. Lipoff in the February 1978 issue of the *IEEE Transactions on Vehicular Technology*.

HF/VHF CRYSTAL OSCILLATOR MODULE

The oscillator whose circuit is given in Figure 5-13 can be operated in the 3-20 MHz range or in the 20-60 MHz range. The collector circuit is broadly tuned to resonance by inductor L and capacitor C. It will deliver a 2-volt RMS signal to a 50-ohm load. It requires 4-9 volts DC for powering it and it draws 20 milliamperes when powered by a 9-volt battery.

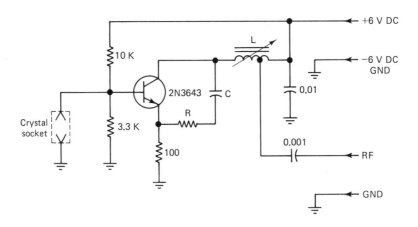

Figure 5-13. HF/VHF Crystal Oscillator Module

HIGH POWER ASTABLE MULTIVIBRATOR

A high power square wave signal source using a single HC2500 IC is an astable multivibrator circuit as illustrated in Figure 5-14. Using the component values noted on the diagram, the no-load output frequency is 8 kHz. The frequency can be changed by using different resistance values for R, R1 and R2, and capacitance for C. With a ± 27-volt power source, this device should be able to deliver a 60-watt output signal. (*Courtesy of RCA Solid State.*)

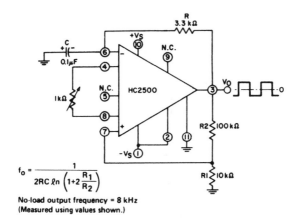

$$f_o = \frac{1}{2RC \, \ell n \left(1+2\dfrac{R_1}{R_2}\right)}$$

No-load output frequency = 8 kHz
(Measured using values shown.)

Figure 5-14. High Power Astable Multivibrator

IC HARTLEY OSCILLATOR

The circuit of a tunable Hartley oscillator utilizing one of the four amplifiers contained in the CA3048 IC is illustrated in Figure 5-15. This circuit can be used as a source of sine wave or square wave signals. When the circuit is powered by a 12-volt DC source, the output signal is a clipped sine wave that has a peak-to-peak level of about 7 volts at the amplifier output. A sine wave signal can be obtained through a coupling capacitor connected to the junction of L1, C1 and C3. C2 should have a large capacitance with respect to C3. (*Courtesy of RCA Solid State.*)

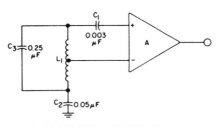

A IS ANY AMPLIFIER OF THE CA3048

Figure 5-15. IC Hartley Oscillator

Nth OVERTONE OSCILLATOR

The oscillator circuit shown in Figure 5-16 may be tuned to the operating frequency by replacing the crystal with a short circuit. When the crystal is cut into the circuit, it will allow the oscillating frequency to be controlled by the mode frequency falling within the tank bandpass. Loop gain may be increased by shunting resistor R_E with a small capacitor. Inductance L_0 may be required to tune out the crystal at the oscillating frequency. The *nth* overtone can represent any number of harmonics or overtones. It is used in lieu of a range of specific numbers. A first overtone is equal to the second harmonic. For example, if a 10-MHz crystal is used, its first overtone and its second harmonic are both 20 MHz. (*Courtesy of CTS Knights Inc.*)

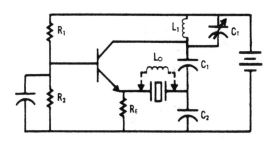

Figure 5-16. Nth Overtone Oscillator

RC PHASE-SHIFT OSCILLATOR

In order for oscillation to occur, feedback from collector to base of a common-emitter circuit must be positive; that is, it must be in phase with the signal at the base. The RC network shown at the collector of transistor Q1 in Figure 5-17 causes the proper phase shift for oscillation.

Output at the collector of Q1 is 180° out of phase with the signal on the base. The network consisting of C1R1, C2R2, and C3R3 shifts the phase of this signal another 180° so that the signal is back in phase when it is reapplied to the base.

If, as is usual, the values for C1, C2 and C3 are all equal, and the values for R1, R2 and R3 are likewise equal, the frequency of oscillation is given by the formula:

$$f_0 = \frac{0.065}{RC}$$

REED-TYPE TONE SQUELCH

A resonant reed is used in the tone squelch circuit given in Figure 5-18 for determining the tone frequency that is generated and to which the decoder responds. When the PTT switch is operated to turn the transmitter on, a tone is generated at the low audio frequency determined by the resonant reed. The tone is amplified by Q4 and is fed through level control R13, C11 and R14 to the transmitter input. In the receive mode, the intercepted tone is fed to terminal 1, through C1 and R1, to the base of Q1 which amplifies

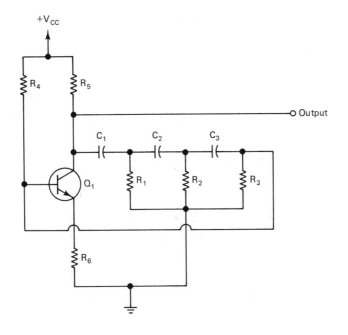

Figure 5-17. RC Phase-Shift Oscillator

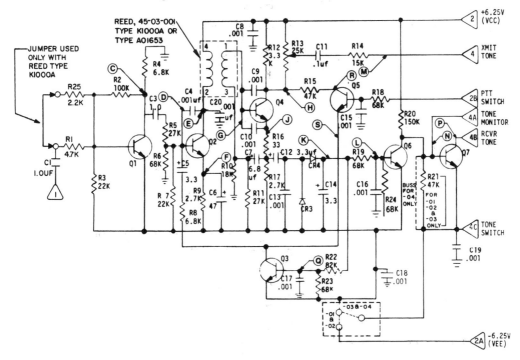

Figure 5-18. Reed-Type Tone Squelch

the tone. Its output is fed through C3 to the base of Q2 through voltage divider R5-R6. The tone, now at a level of 20 millivolts, passes through the primary winding of the resonant reed whose secondary is fed to the base of Q4. (The vibration of the reed induces the tone voltage into the secondary.) The tone at the emitter of Q4 now has a level of 480 millivolts. The tone is passed through R16 to R17. The tone across R17 is fed through C12 to diodes CR3 and CR4, which produce a positive DC output signal that is fed to the base of Q6. Its output is direct coupled to the base of Q7, whose output is fed to the receiver squelch from 4B. (*Courtesy of E. F. Johnson Company.*)

REINARTZ OSCILLATOR

As shown in Figure 5-19, in a Reinartz oscillator positive feedback to sustain oscillation is coupled from a coil (L1) in the collector circuit to a second coil (L2) in the emitter circuit of transistor Q1. Frequency of oscillation is determined by the LC tank consisting of inductor L3 and capacitor C2.

Because the coupling between the frequency-determining L3C2 tank and the transistor is fairly loose, this circuit has good frequency stability. It is often used as a local oscillator in transistor radios.

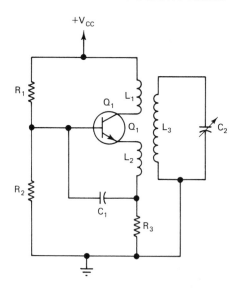

Figure 5-19. Reinartz Oscillator

SCHOCKLEY DIODE OSCILLATOR

A Schockley diode (also called a four-layer diode or a *silicon unilateral switch*) is used in the 10 kHz oscillator shown in Figure 5-20. The capacitor charges until the breakdown voltage of the 2N4984 is reached; then the diode switches on and the inductor causes ringing in the circuit. When current through the 2N4984 drops below the value necessary for holding current, the circuit turns off until the capacitor is again charged up. (*Courtesy of General Electric Company.*)

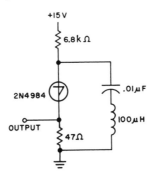

Figure 5-20. Schockley Diode Oscillator

TAPE RECORDER BIAS OSCILLATOR

A push-pull Hartley oscillator circuit is given in Figure 5-21. This oscillator operates at a frequency between 40 kHz and 80 kHz and is used for generating the bias for a magnetic tape recorder. In this circuit, the feedback is from the transistor collector to the opposing transistor base. (*Courtesy of Nortronics Company, Inc.*)

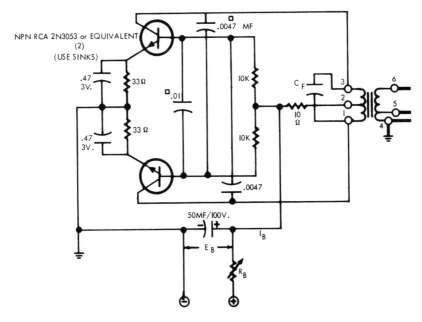

Figure 5-21. Tape Recorder Bias Oscillator

THIRD OVERTONE OSCILLATOR

An overtone oscillator utilizes a crystal that operates at a multiple of its fundamental frequency. Overtone crystals are more economical to use at VHF than fundamental crystals that are ground for VHF operation. The circuit of a third overtone oscillator given in Figure 5-22 utilizes a trap circuit (C_2, L_e, and C_{by}) which must appear inductive to make the crystal operate in the overtone mode. The oscillator frequency is equal to the reciprocal of 2 pi times the square root of $L_e \times C2$. Lo may be required to tune out the crystal at the frequency of oscillation. (*Courtesy of CTS Knights Inc.*)

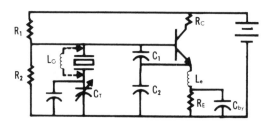

Figure 5-22. Third Overtone Oscillator

TONE BURST ENCODER

A tone burst squelch system is used in some mobile radio applications to activate a mute receiver. Instead of a continuous tone being transmitted, a short burst of tone is generated which modulates the transmitter. The tone is present only at the start of each radio transmission. After the receiver has been activated by a tone burst, the receiver is kept operational by its carrier squelch. When reception of the carrier ceases, the receiver is muted.

In the practical circuit of an actual tone burst encoder given in Figure 5-23, a resonant reed (pin 4 is attached to the collector of Q2) is used for determining and stabilizing the frequency of the tone burst. The tone burst oscillator consists of Q2 and Q4, with the feedback path through the inductive coupling between the primary and secondary windings of the resonant reed assembly. When the PTT switch on the microphone is pressed in to turn the transmitter on, the PTT switch triggers the timing circuit. (*Courtesy of E. F. Johnson Company.*)

TRANSMISSION LINE OSCILLATOR

A quarter-wave section of coaxial cable is used as the parallel resonant circuit of the oscillator circuit given in Figure 5-24. The length of the cable can be calculated by: L (in meters) is equal to 300 divided by the frequency in MHz, times 0.25 times the velocity factor of the coax. For operation at

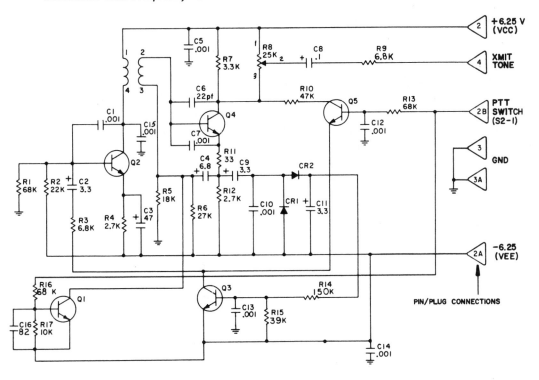

Figure 5-23. Tone Burst Encoder

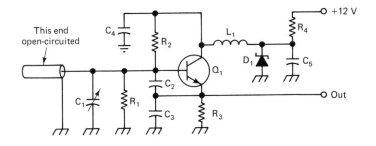

Figure 5-24. Transmission Line Oscillator

460 MHz, for example, 300 divided by 460 equals 0.65 times 0.25 equals 0.1630 times 0.66 (velocity factor) equals 0.1076 meter which can be stated as 10.76 centimeters (4.23 inches).

By leaving the far end of the coax open, it will function as a parallel resonant circuit. By using a voltage regulator to stabilize the collector voltage, fairly good frequency stability is obtained. The variable capacitor at the near end of the coax enables tuning the oscillator over a narrow frequency range.

TUNNEL DIODE OSCILLATOR

The rapid switching and negative resistance characteristics of a tunnel diode can be used to form an oscillator. As shown in Figure 5-25, tunnel diode D1 is connected to a source of positive bias. R1 and R2 are selected to bias the diode in the negative resistance portion of its characteristic curve.

Frequency of this relaxation oscillator is determined by the value of C1 and the characteristics of D1. The capacitor charges until the peak operating voltage of the tunnel diode is reached and operation moves into the negative resistance area of the diode's characteristic curve; then the circuit "relaxes" and C1 discharges.

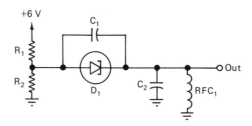

Figure 5-25. Tunnel Diode Oscillator

TWIN-T AUDIO OSCILLATOR

A T-network consisting of two resistors and a capacitor and another T-network consisting of two capacitors and a resistor are used in the twin-T oscillator circuit, shown in Figure 5-26. The oscillating frequency is determined by the values of the resistors and capacitors at which the network causes a 180-degree phase shift. The values noted in the diagram will cause the oscillator to generate a 1 kHz signal. The frequency can be adjusted over a narrow range by making R4 variable.

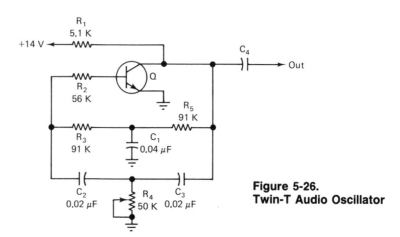

**Figure 5-26.
Twin-T Audio Oscillator**

TWO-TONE AF SIGNAL GENERATOR

It is customary to modulate an SSB transmitter with two audio tones simultaneously when measuring the peak envelope power (PEP) of the transmitter with an oscilloscope. The two audio tone signals must not be harmonically related.

A schematic of a two-tone AF signal generator is given in Figure 5-27. It utilizes a pair of 2N4124 transistors, each functioning as a phase-shift oscillator. Potentiometer R235 is used to balance the output signal to make the levels of the two tones equal.

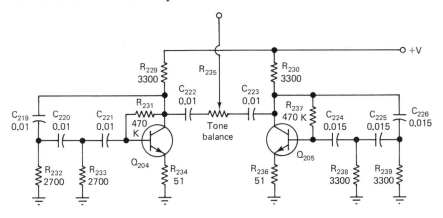

Figure 5-27. Two-Tone AF Signal Generator

VHF HARTLEY OSCILLATOR

A Hartley oscillator circuit using an RCA 40244 NPN transistor is given in Figure 5-28. This circuit was designed for use as the local oscillator of an 88-108 MHz FM broadcast band receiver. The oscillator coil (L4) consists of 3-1/4 turns of No. 18 bare copper wire with an inner diameter of 9/32 inch and a winding length of 5/16 inch. The coil is tapped approximately one turn from its low end. The nominal inductance of this coil is 0.062 microhenry and its unloaded Q is 120. It is tuned through the 77.3-97.3 MHz range by C16, a 6-19.5 pF variable tuning capacitor which is shunted by 2-14 pF trimmer capacitor (C15). The output signal is taken from the top of the tank circuit through C13, a 3-pF NPO capacitor. The transistor is forward-biased by the voltage divider R5-R6, and the collector voltage and bias is obtained from a 15-volt DC power source. (*Courtesy of RCA Solid State.*)

VOLTAGE CONTROLLED OSCILLATOR

The circuit of a voltage controlled oscillator utilizing an SP1648B IC is given in Figure 5-29. It utilizes two varactor diodes for varying its frequency.

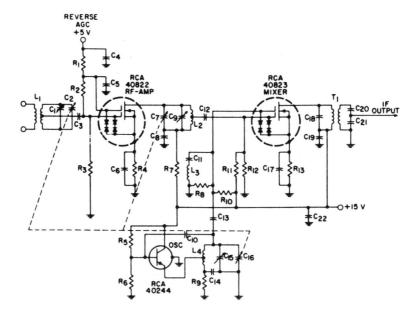

Figure 5-28. VHF Hartley Oscillator

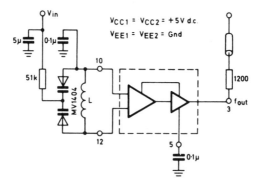

Figure 5-29. Voltage Controlled Oscillator

Using a 5-turn coil of No. 20 enameled copper wire wound on a T30-13 Micro Metal toroid core, the frequency can be adjusted from 50 to 170 MHz by feeding 2 to 10 volts DC to the junction of the two varactor diodes. Although not shown in the diagram, the pins of this 14-pin dual-in-line IC are connected as follows: pins 1 and 14 to + 5 volts DC, pins 7 and 8 to the negative side of the supply voltage, pin 10 to one end of the coil and to the anode of one of the varactors, pin 12 to the other end of coil L and the anode of the other varactor, pin 5 to ground through an 0.1-μF capacitor and pin 3 to the output load. (*Copyright Plessey Semiconductors.*)

WIEN BRIDGE OSCILLATOR

Figure 5-30 illustrates a Wien bridge oscillator utilizing a single operational amplifier. The outstanding feature of the Wien bridge circuit is that two adjacent arms of the bridge contain, respectively, a parallel RC network and a series RC circuit. Phase shifting through the bridge causes the output of the oscillator to be fed back in phase to the input of the op amp. If $R_f = 2R_i$, frequency of oscillation of this circuit is given by the formula:

$$f = \frac{1}{2\pi RC}$$

(Courtesy of Opamp Labs Inc.)

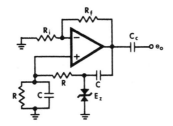

Figure 5-30. Wien Bridge Oscillator

CHAPTER 6

Mixers and Frequency Converters

INTRODUCTION

A mixer produces a single output signal from two or more input signals. One type of mixer is a simple voltage adding circuit, such as is used in an audio mixer. Here, the levels of several different audio signals are controlled and blended into a single audio sound; in this manner it is even possible to electronically "marry" a vocalist with background music, horns with guitars, etc.

A mixer of quite a different stripe is the *heterodyne* mixer used in RF and other electronic circuits. In this type of circuit, two frequencies together are used to produce a third frequency. By combining RF and oscillator frequencies, for example, a heterodyne mixer produces the intermediate frequency (IF) signal used in a radio receiver.

The heterodyne mixer utilizes the principle of heterodyne action. This occurs when two alternating voltage signals of different frequencies are "beat" together electronically. Many different signals result from this mixing, but two of the strongest, and generally the most useful, are: 1) the *sum* of the two original frequencies, and 2) the *difference* of the two original frequencies. Other components produced at the mixer output include both of the original signal frequencies, harmonics (that is, multiples) of these frequencies, harmonics of the beat frequencies, etc.

A down-converter is used for converting an RF signal to a lower frequency. An up-converter converts an RF signal to a higher frequency.

Although the harmonics produced by mixing are sometimes used, a conventional AM superheterodyne receiver provides a good example of using mixing to produce an intermediate frequency.

As shown in Figure 6-1, the RF stages are tuned to 1450 kHz, and the RF signal containing the modulated intelligence is fed to the mixer. If this is mixed with a local oscillator frequency of 1905 kHz, the components at the output of the mixer will include: the RF signal frequency (1450 kHz); the local oscillator frequency (1905 kHz); the sum of the RF and LO frequencies (3355 kHz); the difference of the RF and LO frequencies (455 kHz); and many other combinations produced by the fundamental frequencies and harmonics of the LO and RF.

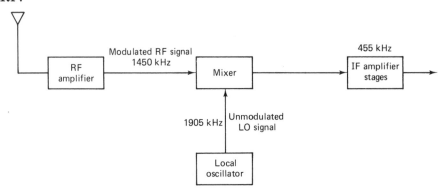

Figure 6-1. Heterodyne Mixing

In an AM radio, the difference is usually used for the IF; therefore, the IF stages are narrowly tuned to 455 kHz, causing them to reject the other frequencies available. The 455 kHz produced contains the modulation that will later be detected. In the case of the superheterodyne receiver of this example, the local oscillator must track the incoming signal frequency, so the LO and RF tuning capacitors are mechanically "ganged" together.

At higher radio and microwave frequencies, noise and spurious or unwanted harmonics can be a problem in mixing circuitry. Sometimes a balanced mixer is used to reduce spurious harmonics and to allow low-level input signals to be applied. Also, components such as hot-carrier diodes are often used because of their low noise characteristics at high frequencies.

In common parlance, the word *converter* is used to describe a circuit that functions both as an oscillator and a mixer. Frequency conversion can be performed by mixing the oscillator signal and another signal together. Another method of frequency conversion involves generating harmonics of an input signal, then tuning the output to the desired harmonic.

This chapter illustrates a number of simple and more complex circuits for mixing and frequency conversion.

ACTIVE BALANCED JFET MIXER

The circuit of an active balanced mixer using two U310 JEFTs is shown in Figure 6-2. The input signal is fed in through the 50-ohm input and capacitor C1 at the left of the circuit, while the local oscillator signal is coupled through C5. Variable capacitors C2 and C4 are used to balance the incoming signals to the proper levels. Output transformer T1 can be a RELCOM type BT-9 for the 150 MHz band. The noise figure of the circuit shown is around 8 dB. (© *Siliconix incorporated.*)

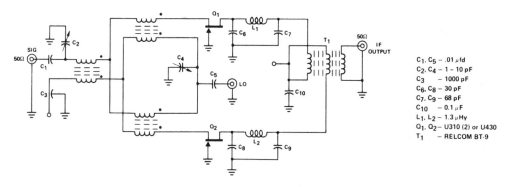

Figure 6-2. Active Balanced JFET Mixer

AUDIO MIXER AND PREAMPLIFIER

The audio mixer circuit given in Figure 6-3(a) is a voltage adding circuit, and is quite different from a heterodyne mixer. It contains four preamplifiers and two output amplifiers which can be used to feed the output into a mono or stereo sound system. Each microphone jack automatically grounds the input of the microphone preamplifier to minimize noise and hum pickup whenever the mike plug is removed from the jack. In Channel 1, for example, when a mike is plugged in, its output is fed through a 620-ohm resistor and a 1-uF capacitor to the base of TR1, an NPN transistor which is direct-coupled to TR2. The output of TR2 is fed to the base of TR3 which is direct-coupled to TR4. The Channel 1 and Channel 2 microphone circuits are identical to each other. The Channel 1 phono and the Channel 2 phono input circuits are almost identical to each other. The Channel 1 and Channel 2 auxiliary input circuits are identical to each other. The specifications on this mixer are listed in Figure 6-3(b).

When set in the mono mode with switch SW2, the Channel 1 inputs are fed to the output amplifiers shown at the top right of the diagram and also the output amplifier for Channel 2. Overall volume is controlled with the pot at the input of the Channel 1 output amplifier. Individual level controls are provided for each of the preamplifier outputs. When SW2 is in the stereo mode, the Channel 1 circuits remain as above, but the Channel 2 inputs are fed to the Channel 2 output amplifier shown at the center right of the diagram, through the Channel 2 master volume control.

Phono inputs are connected to the input of a mike preamplifier when the phono switch is set in the phono position. The auxiliary input of each channel is connected through a voltage divider to the master volume control of each channel. The power supply utilizes a full-wave bridge rectifier and a pass transistor voltage regulator. (*Courtesy of Radio Shack, a Division of Tandy Corp.*)

AUTODYNE CONVERTER

The autodyne converter is used in radios for converting a radio-frequency input signal to an intermediate frequency for application to the IF amplifier. A single transistor is used as both oscillator and mixer in the autodyne circuit.

As shown in Figure 6-4, the proper RF input signal is selected by tuning capacitor C1, and is coupled to the base of transistor Q1. C1 is mechanically ganged to capacitor C3 in the oscillator portion of the circuit.

Oscillation occurs when a harmonic of the signal at the collector is coupled through the upper winding of transformer T2 to the lower winding, tuned to the proper frequency by C3, and coupled back to the emitter as regenerative feedback.

The RF input signal is mixed with this local oscillator frequency in transistor Q1. The output of Q1 is the IF frequency, and is developed across the tuned circuit consisting of C5 and the primary winding of T3.

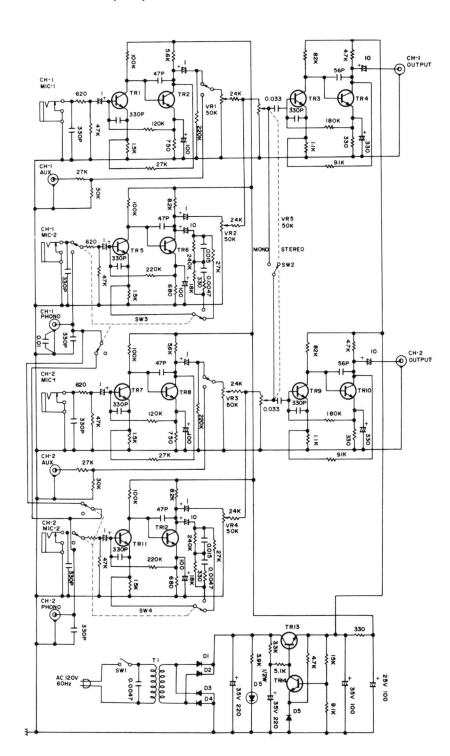

Figure 6-3(a). Audio Mixer and Preamplifier

SPECIFICATIONS

MIC 1-2 Input Sensitivity:	1.0 mV	(Out = 0.775V)
AUX Input Sensitivity:	90 mV	
PHONO Input Sensitivity:	1.6 mV	(Out = 0.775V)
Distortion (30-20kHz):		
MIC 1-2	0.5%	(Out = 0.775V)
PHONO	0.5%	(Out = 0.775V)
AUX	0.5%	(Out = 0.775V)
Signal-to-Noise Ratio:		
MIC 1-2 (with shorted input)	50 dB	(Out = 0.775V)
PHONO (with shorted input)	55 dB	(Out = 0.775V)
AUX (with shorted input)	55 dB	(Out = 0.775V)
Signal-to-Hum Ratio:	60 dB	
Maximum Output Level:		
MIC-PHONO-AUX (1 kHz)	7.5 Volts	Distortion: Under 10%
Residual Noise:	0.5 mV	
Frequency Response:	20-20 kHz	
	RIAA $\pm$ 3 dB	
MIC Over-Load:	350 mV	
PHONO Over-Load:	350 mV	
Power Requirements:	120 volt 60 Hz (4 watts)	
	(240V 50Hz for Australian models)	

Figure 6-3(b). Specifications

BALANCED RF MIXER

Two diodes are used in the balanced mixer circuit given in Figure 6-5. Intercepted radio signals, after being translated to the receiver IF, are inductively coupled to the secondary of the left-hand transformer. This is a center-tapped coil that is tuned to the translated radio signal frequency with the variable capacitor that is shunted across it. The two diodes are connected in series-aiding. The LO signal is fed in at the center tap of the secondary of the input transformer. The output transformer secondary is tuned to the IF signal frequency. (*Courtesy of RCA Solid State.*)

CLASS C FREQUENCY TRIPLER

Class C operation and the principle of a "ringing" tank circuit can produce frequency multiplication. As shown in Figure 6-6(a), the schematic diagram of a basic multiplier is almost identical to that of a class C RF amplifier. In the multiplier, however, the output tank is tuned to a harmonic (in this case, the third harmonic) of the input frequency.

Being biased for class C operation, transistor Q1 turns on and draws collector current during only a small portion of each input cycle. This short pulse of collector current causes the ouput tank to oscillate, or "ring," at a harmonic frequency that is three times the original input frequency. Because

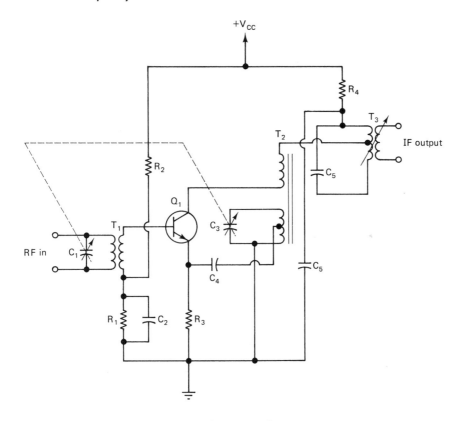

Figure 6-4. Autodyne Converter

the pulse of collector current is applied only once for each three cycles of output, there is some damping of the output amplitude between input pulses.

Like the class C amplifier, this type of circuit is very efficient because the driving element is turned on during only a small portion of the input cycle.

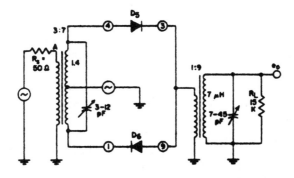

Figure 6-5. Balanced RF Mixer

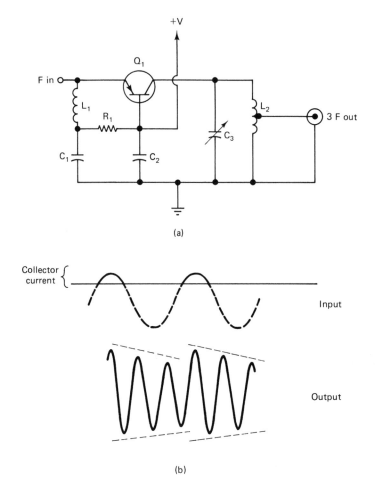

(a)

(b)

Figure 6-6. Class C Frequency Tripler

CRYSTAL-CONTROLLED DOWN-CONVERTER

The circuit of a mixer and a crystal-controlled oscillator, which is given in Figure 6-7, is typical of the circuitry used in FM communications receivers. This circuit is designed to follow the first down-converter of a double-conversion superheterodyne receiver. In a typical receiver, the input signal is at 10.7-MHz. This signal is passed through the crystal filter which limits the bandwidth to the base of transistor Q601, which with Q602 forms a 10.7-MHz IF amplifier. The output of this amplifier is tuned to 10.7 MHz with L602, and the output is fed from the junction of C609 and C610 to the input of the second mixer, IC601. The output of the crystal-controlled oscillator, Q603, which generates a 10.245-MHz signal, is also fed to the mixer from the junction of C614 and C615. The output of the mixer is fed through T600 and a 455-kHz bandpass filter to an IF amplifier (not shown).

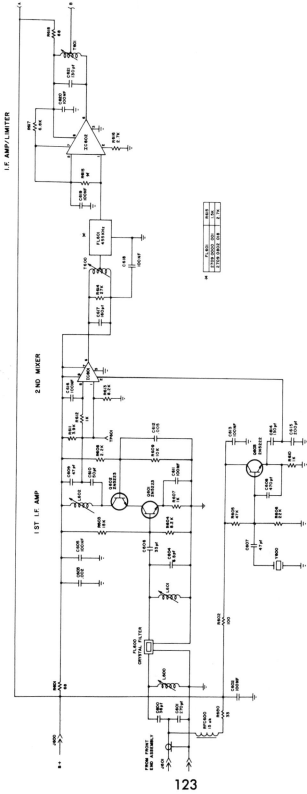

Figure 6-7. Crystal-Controlled Down-Converter

C600-C601, C609-C610 and C614-C615 form voltage division and impedance matching networks. With the exception of C614-C615, they also are part of parallel resonant circuits. All provide impedance matching and voltage reduction. (*Courtesy of Aerotron, Inc.*)

DIODE MIXER

A simple diode mixer using a single diode is illustrated in Figure 6-8. The input tank, consisting of capacitor C1 and the primary winding of transformer T1, is tuned to frequency f1; C2 and T2 are tuned to frequency f2. The mixed frequency is developed across C3 and the primary winding of T3, and is coupled through transformer T3 to the output. With the proper bias applied, a tunnel diode can be used for CR1.

Diode mixers introduce very little noise, and are in wide use, especially for applications above 400 MHz.

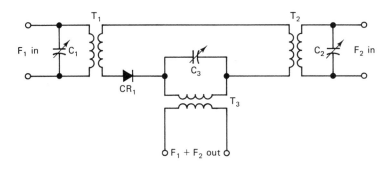

Figure 6-8. Diode Mixer

DOUBLE BALANCED MIXER

A double balanced mixer provides high isolation between the local oscillator and an intercepted RF signal. The circuit given in Figure 6-9 is of a Watkins-Johnson WJ-M1A-11 mixer which utilizes four diodes, is operable with LO and RF signals within the 3-1300 MHz range, and can be used to generate an IF signal in the DC to 1200-MHz range. Conversion loss and noise figure in SSB applications are both 7.5 dB when all signals are within the 10-100 MHz range. Conversion loss rises to 10 dB at frequencies below 10 MHz and above 100 MHz. Access to the three ports is via BNC connectors. All inputs and the output are designed for 50 ohms. The LO should deliver a + 7 dBm signal to the LO port. The RF signal is fed to the R port and the IF signal is taken from the I port. (*Courtesy of Watkins-Johnson Company.*)

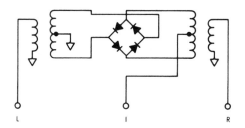

Figure 6-9. Double Balanced Mixer

DUAL GATE MOSFET MIXER

In the mixer circuit given in Figure 6-10, the local oscillator signal is fed to gate 2 of a dual gate MOSFET. This gate is forward biased by a positive DC voltage obtained from the junction of voltage divider R2-R3. The incoming RF signal is fed through C2 to gate 1 which is reverse biased by the voltage drop across the drain resistor. The output tank circuit can be tuned to either the sum or difference beat frequency of the RF signal and LO frequencies.

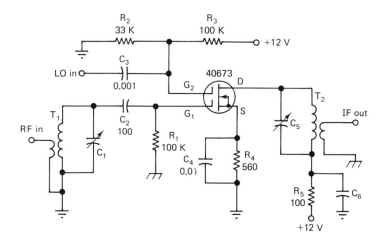

Figure 6-10. Dual Gate MOSFET Mixer

FM RECEIVER FRONT END UTILIZING A MOSFET RF AMPLIFIER

A single gate MOSFET is used as the RF amplifier in the FM receiver front end circuit given in Figure 6-11. The MOSFET output is fed from a tap on L2 and through C9 to the base of the mixer transistor. The local

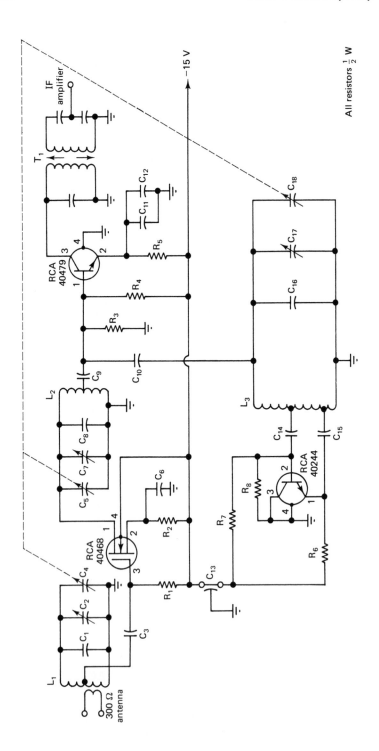

Figure 6-11. FM Receiver Front End Utilizing a MOSFET RF Amplifier

oscillator employs an NPN transistor whose output is injected into the mixer base through C10. The tuner provides 34.5 dB of gain, of which about a third is provided by the MOSFET RF amplifier and the rest by the mixer. (*Courtesy of RCA Solid State.*)

FOUR-CHANNEL AUDIO MIXER

A single CA3048 IC is used to provide gain and isolation of four audio channels in the voltage adder mixer circuit given in Figure 6-12. The CA3048 IC contains four independent amplifiers, each with a differential input. Each amplifier contained in this 16-pin dual inline IC has an open loop gain of 58 dB and an input impedance of 90,000 ohms. In the mixer circuit, each input is fed to the noninverting input of its associated amplifier through a 500,000-ohm level control potentiometer. Resistors R5, R6, R7 and R8 are used to determine the gain of the amplifier. With the values shown in the diagram, 20 dB of gain is provided for each channel. The gain may be increased by using a lower value of feedback resistor. By using a 1000-ohm resistor in lieu of an 820-ohm resistor, gain can be increased to around 34 dB. A DC power source delivering + 12 volts to pin 15 of the IC can be used to power the mixer. The RC network at the output of each amplifier stabilizes the amplifiers when the sources and load do not provide adequate damping. The output of the mixer may be connected to a load of at least 10,000-ohms. (*Courtesy of RCA Solid State.*)

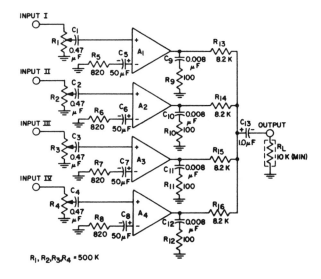

Figure 6-12. Four-Channel Audio Mixer

FREQUENCY DIVIDER CHAIN

Sawtooth pulses from a 2N2647 unijunction transistor oscillator are divided in frequency by each succeeding stage of the frequency divider chain shown in Figure 6-13. Each stage employs a four-layer diode (also called a silicon unilateral switch, or SUS); the output from each stage is one-half the frequency of the preceding stage output. (*Courtesy of General Electric Company.*)

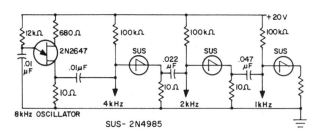

Figure 6-13. Frequency Divider Chain

HALF-WAVE DOUBLE BALANCED MIXER

The circuit given in Figure 6-14 is of a half-wave double balanced mixer utilizing a pair of VMP-4 power MOSFETs. The intercepted radio signal is coupled through T2. The local oscillator signal is coupled through T1 and the resulting IF signal is taken out through T3. (© *Siliconix incorporated.*)

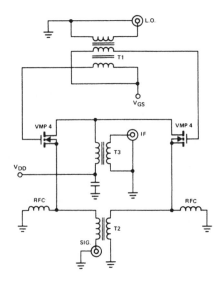

Figure 6-14. Half-Wave Double Balanced Mixer

MIXING TRANSFORMER

Transformers with as many as six 600-ohm windings are available for mixing. Signals that are fed to any of the windings will be present at all of the other windings. An application for the Audisar 9K600-6 transformer is illustrated in Figure 6-15, an impedance-matching adder and signal splitter. Here a telephone line is fed to winding 1 for feeding a program from a remote location into the system. Outputs of windings 3 and 4 are fed to two different sound systems. Winding 6 is a spare. (*Courtesy of Audisar.*)

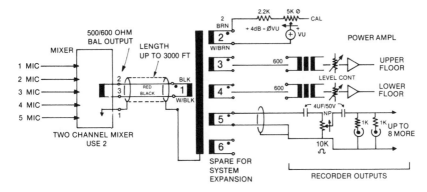

Figure 6-15. Mixing Transformer

MOSFET MIXER FOR 10-METER BAND RECEIVER

The circuit of a low-noise mixer for use in a 28-MHz band receiver is given in Figure 6-16. It employs a dual-gate TIS148 MOSFET. The 28-MHz radio signal is capacitively coupled to gate 1 and the local oscillator signal is capacitively coupled to gate 2. The output signal is capacitively coupled to the input of a 10.7-MHz IF amplifier. The input is tuned by varying the inductance of L1 and the output is tuned to 10.7 MHz by adjusting L2. The local oscillator signal may be crystal-controlled and its frequency should be 10.7 MHz higher than the intercepted signal frequency. Or the LO may be tunable through the 38.7-40.4 MHz range to enable tuning through the 28-29.7 MHz band. (*Courtesy of Texas Instruments.*)

PUSH-PULL MIXER

The circuit given in Figure 6-17 is of a push-pull mixer utilizing a pair of NPN transistors. The LO signal is injected into the bases of Q1 and Q2 through C3 and the center tap of the secondary of T1. C4 and C3 form a capacitive voltage divider that reduces the level of the LO signal. T1 is tuned to the incoming signal frequency by adjusting the core (not shown) of the RF transformer. The IF transformer T2 is tuned in the same manner to the IF signal frequency.

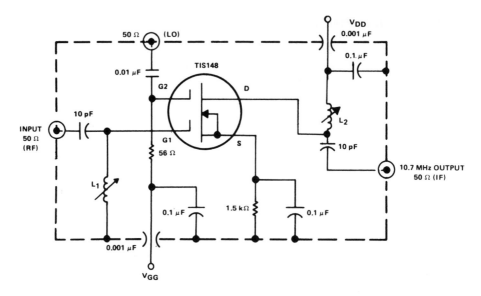

Figure 6-16. MOSFET Mixer for 10-Meter Band Receiver

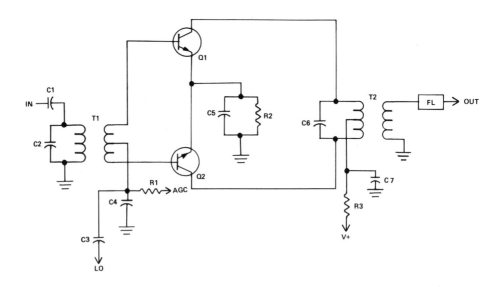

Figure 6-17. Push-Pull Mixer

PUSH-PUSH DOUBLER

Figure 6-18 shows a frequency doubler that uses a pair of NPN transistors connected for push-push operation. This type of circuit is frequently used in VHF and UHF radio equipment.

Transistors Q1 and Q2 are connected as common-emitter amplifiers, and they are biased for class C operation by resistors R1 and R2. In the collector circuit is a tank, consisting of capacitor C2 and the primary winding of transformer T2, that is tuned to the 2nd harmonic of fundamental frequency F1.

The input at the fundamental frequency causes out-of-phase signals to appear at the bases of the two transistors. When the input at the base of Q1 swings positive, that transistor is turned on for a portion of the positive half-cycle, and a surge of collector current causes "ringing" in the collector tank circuit. At the same time, Q2 is cut off by the negative swing of the input to its base.

During a portion of the opposite half-cycle of input, Q2 is turned on by a positive potential on its base, and collector current through Q2 again causes "ringing" in the C2T2 tank. Q1 is cut off at this time.

Since each transistor is biased for class C operation, there is a time during each cycle of input when neither Q1 nor Q2 is turned on, and the "ringing," or oscillation, of the collector tank circuit causes an output at twice the fundamental frequency. The push-push configuration is highly efficient and can be used to produce an output at even harmonics—as a doubler, quadrupler, etc.

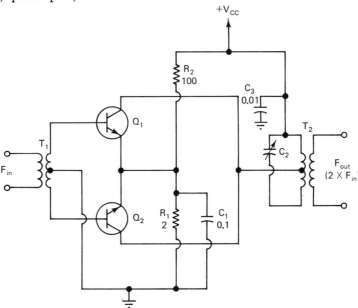

Figure 6-18. Push-Push Doubler

RECEIVER FREQUENCY MULTIPLIER

Essentially all VHF and UHF receivers employ frequency multipliers to obtain a local oscillator injection frequency that is adequately high to

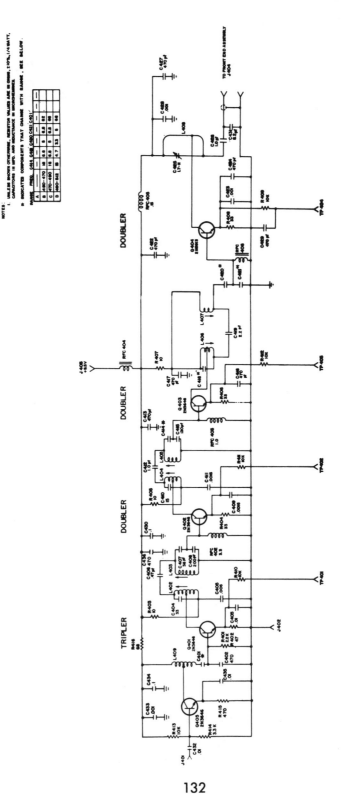

Figure 6-19. Receiver Frequency Multiplier

132

enable use of relatively low frequency crystals in the oscillator. An example of a receiver frequency multiplier circuit is given in Figure 6-19. To operate a UHF receiver on 462.55 MHz, an 18.827-MHz crystal is required since the multiplier utilizes a tripler and three cascaded doublers to provide frequency multiplication of 24 times, and because the mixer injection frequency must be 10.7 MHz lower than the operating frequency. The crystal frequency of 18.827 MHz is multiplied 24 times to 451.848 MHz. When heterodyned with an intercepted signal on 462.55 MHz, an IF signal at 10.702 MHz will be generated. This 2-kHz frequency error can be corrected by rubbering the crystal, by adjusting a capacitor in series shunted across the crystal, or by adjusting a variable inductor in series with the crystal. (*Courtesy of Aerotron, Inc.*)

SINGLE BALANCED MOSFET MIXER

A pair of dual gate MOSFETs are shown in the single balanced mixer circuit given in Figure 6-20. The input signal is fed in through T1 which is used as a balanced driver. The output is also balanced by T2. Both T1 and T2 are baluns in a sense, since they interface a balanced circuit with an unbalanced circuit.

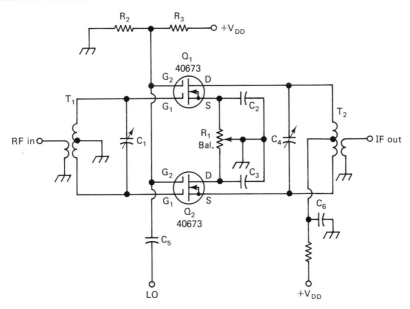

Figure 6-20. Single Balanced MOSFET Mixer

TRANSISTOR MIXER

A simple additive mixer using a single NPN transistor is illustrated in Figure 6-21. Frequency f1 is developed across the input tank consisting of capacitor C1 and the primary winding of impedance-matching transformer T1. It is then applied to the base of transistor Q1 through capacitor C4, which isolates T1 from the DC bias used for the transistor. Frequency f2 is applied through T3 and C5 to the emitter of the same transistor. The tank in the collector circuit of Q1 is tuned to the sum (or the desired harmonic) of the two input frequencies.

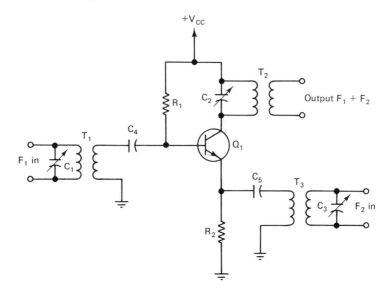

Figure 6-21. Transistor Mixer

TRANSMITTER FREQUENCY MULTIPLIER

Most FM communications transmitters utilize a chain of frequency multiplier stages to multiply the oscillator frequency to the assigned transmitter carrier frequency and to multiply the FM deviation to the maximum permitted value. In the transmitter frequency multiplier circuit given in Figure 6-22, the frequency and FM deviation are multiplied 36 times. For example, if the transmitter is authorized to operate on 467.55 MHz and frequency deviation of ± 5 kHz is authorized, the oscillator must be fitted with a crystal for operation on 12.9875 MHz. This 12.9875-MHz signal is tripled by Q202 to 38.9625, tripled again by Q203 to 116.8875, doubled by Q204 to 233.775 MHz and doubled again by Q205 to 467.550 MHz. If the maximum FM deviation at the input of the first tripler is ± 138 Hz, the FM deviation will be multiplied ± 4968 Hz. The output of Q205 is fed through a bandpass filter that attenuates unwanted harmonics and all unwanted signals.

The output of each frequency multiplier stage is tuned to the second (doubler) or third (tripler) harmonic of its input signal. All of the transistors

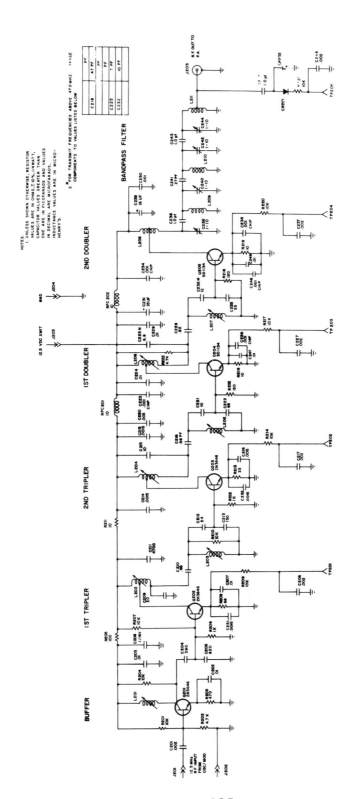

Figure 6-22. Transmitter Frequency Multiplier

135

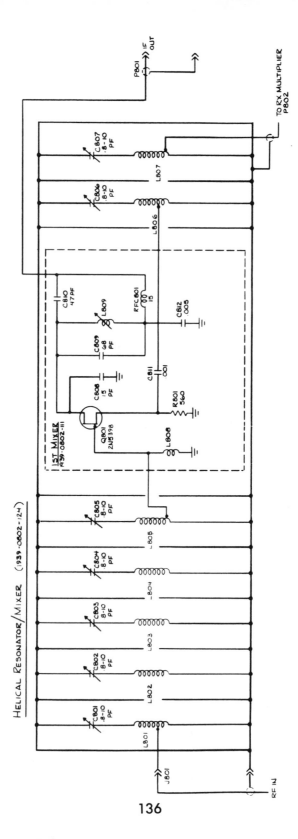

Figure 6-23. UHF FET Mixer

used as frequency multipliers are biased for class C operation. Collector current flows only during the positive signal excursions of the driver signal, and then for only part of each cycle. The output tank of each multiplier is tuned to the required harmonic. Because of the flywheel effect of the tank circuit, the output signal is converted into a reasonable facsimile of a sine wave. (*Courtesy of Aerotron, Inc.*)

UHF FET MIXER

A 3N5398 FET is used as the mixer in a 450-512 MHz band FM receiver in the front end circuit given in Figure 6-23. The intercepted radio signals are fed to a tap on L801 which is tuned with C801. Coils L801, L802, L803, L804 and L805 represent a helical resonator which is tuned to pass a band of frequencies. The gate of the FET is fed from a tap on L805 which is shunted by L808, an air wound coil consisting of 7 turns of No. 28 wire with an inside diameter of 0.12 inch.

The frequency-multiplied LO signal is fed to the tap on L807 and is taken from a tap on L806 and injected through C811 into the source of the FET. The FET drain load is variable inductor L809 which is tuned to the IF signal frequency. The IF signal is coupled through C810 to the input of the IF amplifier (not shown). The two-section helical resonator which is used in the LO injection circuit removes the harmonics and other spurious signals which could cause the receiver to intercept unwanted signals. (*Courtesy of Quintron Corporation.*)

VARACTOR FREQUENCY DOUBLER

The varactor diode has excellent harmonic generation, especially at high frequencies, and this feature is used to create the frequency doubler shown in Figure 6-24.

An input series resonant filter, consisting of capacitor C2 and inductor L1, is tuned to the input frequency. The capacitive reactance of varactor diode D1 varies depending on the input voltage; the varactor is biased to produce generous harmonics at the input frequency. At the output of the circuit, a second series resonant tank is tuned to the second harmonic of the

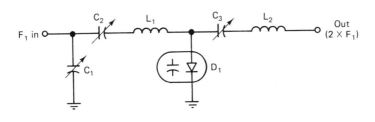

Figure 6-24. Varactor Frequency Doubler

input. This filter rejects the fundamental frequency and all other harmonics. With a 10-watt input, a varactor doubler can produce an output of 5 watts at twice the input frequency.

VARACTOR FREQUENCY TRIPLER

A basic frequency tripler employing a varactor diode is shown in Figure 6-25. At the input to varactor diode D1, a series-resonant circuit consisting of capacitor C1 and the secondary winding of transformer T1 is tuned to the input frequency. L1 and C2 form an "idler" resonance, tuned to the second harmonic of the input. At the output, C3 and the primary of T2 are tuned to the third harmonic (three times the input frequency). R1 is a high resistance (around 100 kilohms) used for varactor bias.

This type of circuit is commonly used at very high radio frequencies. At even higher frequencies, a step recovery diode can be substituted for the varactor. When a step recovery diode is used, no "idler" is needed.

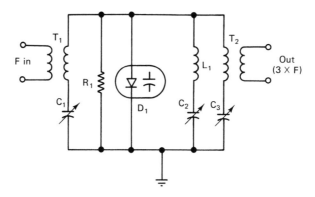

Figure 6-25. Varactor Frequency Tripler

CHAPTER 7

Signal Conditioning
and
Interface Circuits
and Devices

INTRODUCTION

In addition to amplification, frequency conversion, modulation and demodulation, electrical signals are subjected to other kinds of conditioning, including changing impedance, attenuating signal levels, filtering out unwanted frequencies, adjusting the frequency response, etc. Interface devices are used for matching a balanced circuit to an unbalanced circuit, a high impedance source to a low impedance load, and feeding a low impedance load from a high impedance source.

139

Interface devices are sometimes passive; they use no power-consuming devices such as transistors. Some examples of passive devices are the passive filter and the interface that allows two speakers to be served by a single amplifier output.

Other conditioning and interface circuits are active; that is, they amplify the signal in addition to altering it. Some of these active devices include the active filter, op amp integrator, PCM codec, etc. This chapter covers a large number of signal conditioning and interface circuits and devices.

AUDIO AMPLIFIER OUTPUT SPLITTERS AND COMBINERS

A transformer known as the Mix-N-Match used to be marketed by Alco Products which has dropped the transformer from its product line. However, equivalent transformers can be ordered custom-made by a transformer manufacturer such as Staco Energy Corp., which has offered to make these transformers in small quantities.

A 3-winding transformer can be used to split the output of an audio amplifier between two speakers as shown in Figure 7-1(a). The transformer introduces a minimum loss of 3 dB. Or, a 3-winding transformer can be used as an audio signal combiner in such an application as, for example, converting a stereo signal into a mono signal, as shown in Figure 7-1(b). Or the signal combiner can be used to feed the output of a CB radio and a scanner receiver to the same speaker for simultaneous monitoring.

The Mix-N-Match transformers had three 8-ohm windings and were available in 5-watt and 50-watt power ratings.

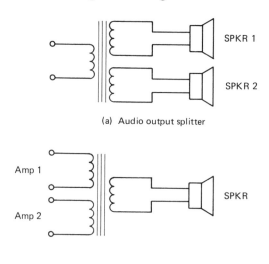

(a) Audio output splitter

(b) Audio output combiner

Figure 7-1. Audio Amplifier Output Splitters and Combiners

COAXIAL RELAY CONTACT CONFIGURATIONS

Coaxial relays usually have their contacts enclosed within a metal cavity to provide shielding and to prevent impedance discontinuities. They are available in the various contact configurations shown in Figure 7-2. Ground contacts are used in relays, which, when energized, ground the transmitter output. Non-grounding contacts are used in relays for circuits where the disengaged contact blade does not ground against the cavity wall. Resistor-terminated circuits are used to ground the engaged connector through a resistor. (*Courtesy of Magnecraft Electric Company.*)

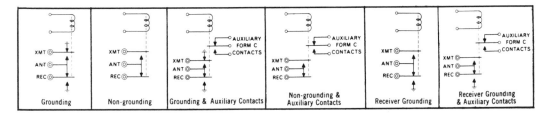

Figure 7-2. Coaxial Relay Contact Configurations

DIFFERENTIATOR

A differentiator is used for converting a square wave into a narrow pulse. The passive differentiator circuit shown in Figure 7-3(a) consists of an R-C network. Capacitor C is in series with the signal path and the resistor (R) is shunted across the output of the network. C charges almost instantly to the input signal voltage level and R dissipates the charge so that C will be ready to accept charges when the next signal cycle begins.

An active solid state differentiator circuit is given in Figure 7-3(b). It utilizes a CA3008 operational amplifier IC. The 51-ohm resistor in series with the input signal limits the high frequency numerical gain factor to 433. Figure 7-3(c) shows the input and output waveform at 1 kHz. (*Courtesy of RCA Solid State.*)

DIODE CLAMPER

A clamping circuit is used to clamp a waveform to a particular DC axis. As Figure 7-4 illustrates, a clamper can be created from a capacitor in series and a diode and resistor in parallel. The circuit clamps an unsymmetrical waveform at the input to the 0-volt axis line at the output. For proper operation, the RC time constant should be greater than five times the period of the input wave; that is, $\frac{5}{f} < RC$.

With the addition of a DC potential in series with diode CR1, the output can be clamped to values other than zero.

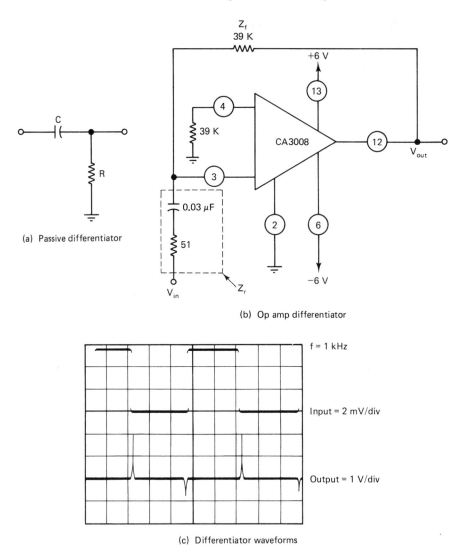

(a) Passive differentiator

(b) Op amp differentiator

(c) Differentiator waveforms

Figure 7-3. Differentiator

DIODE CLIPPER

One of the basic uses of a diode is as a *clipper*, to clip or remove the unwanted portions of a waveform. The simplest clipper consists of a fairly large (about 10 kilohm) resistor in series and a diode in parallel, as shown in Figure 7-5(a). During that portion of input when the diode is forward-biased, it shorts the input to ground. During the opposite half-cycle, the diode does not conduct, and the input is unaffected.

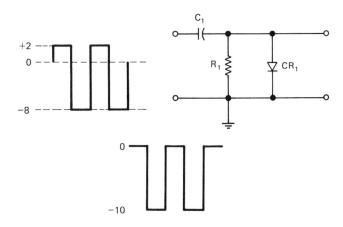

Figure 7-4. Diode Clamper

Figure 7-5(b) illustrates how a parallel diode can be biased to clip only a *portion* of the incoming waveform. By biasing the diode with a 5 VDC source, the circuit now clips off only that portion of input that is above +5 volts.

A clipper can also be constructed using a diode in series and a resistor in parallel, as shown in Figure 7-5(c).

FET PHASE INVERTERS

Two similar FETS are used in the phase inverter circuit given in Figure 7-6(a). Assume that when the input signal at the gate (G) of Q1 is negative-going, the amplified signal at point X will be positive-going because of the phase inversion through Q1. At the same instant, the weaker signal at point Y (junction of R5-R6 voltage divider) is also positive-going and is fed to the gate of Q2. The amplified signal at point Z will be negative-going because of the phase inversion through Q2. The gates of Q1 and Q2 are driven in opposite phase. Hence, their output signals will be opposite in phase and are fed to the following push-pull stage.

A single FET can be used as a phase inverter in the circuit given in Figure 7-6(b). When the input signal is negative-going at point X, the output signal at point Z (source of Q1) will be negative-going since the signal at the source(s) is in phase with the signal at the gate. Thus, the signals fed to the following push-pull stage are of opposite phase.

Equal and opposite-phase signals will appear at points Y and Z when the resistance of R2 (drain load) is equal to the sum of the resistances of R3 and R4 (source load) since drain (D) and source current are the same. The FET is reverse-biased only by the voltage drop across R3 since the gate return resistor (R1) is connected to the junction of R3 and R4.

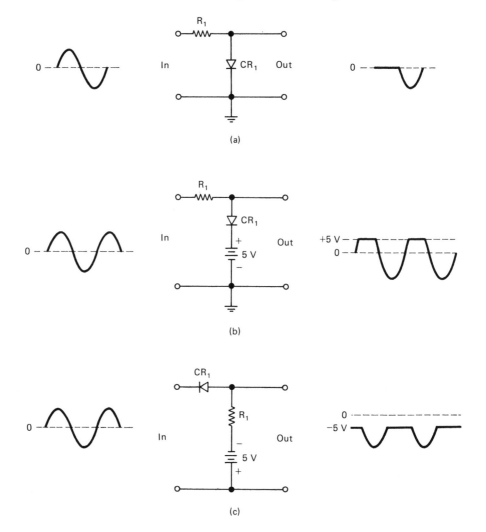

Figure 7-5. Diode Clipper

FM TRANSMITTER ROLLOFF FILTER

Although pre-emphasis is used in FM transmitters to increase the FM deviation at the higher frequencies, it is also customary to attenuate the high audio frequencies to limit the bandwidth occupied by the signal. A schematic of a de-emphasis circuit and a rolloff filter is given in Figure 7-7. This circuit is inserted between the output of the modulation limiter and the input of the phase modulator or frequency modulator. When digital modulation is used, a rolloff filter is not required. (*Courtesy of Heath Company.*)

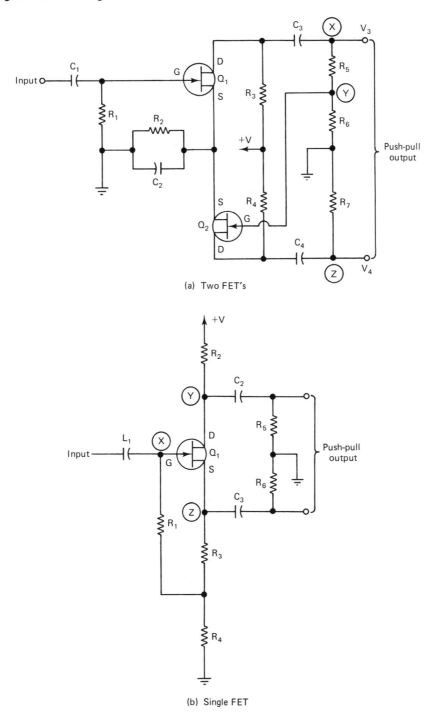

(a) Two FET's

(b) Single FET

Figure 7-6. FET Phase Inverters

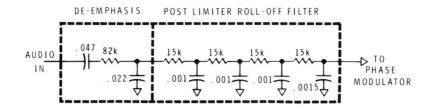

Figure 7-7. FM Transmitter Rolloff Filter

FREQUENCY-TO-VOLTAGE CONVERTER

Figure 7-8 shows a frequency-to-voltage converter that accepts almost any waveform in the DC-10kHz range and yields a positive output in proportion. The components inside the dotted lines are all part of a Teledyne 19400 IC. Besides the voltage output, this circuit also provides a buffered output at the same frequency as the input, as well as a frequency that is one-half the input frequency. The offset adjustment potentiometer is used to set the output voltage to zero with a DC input. (*Courtesy of Teledyne Semiconductor.*)

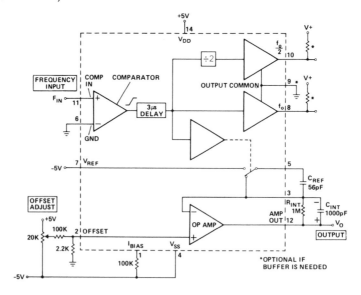

Figure 7-8. Frequency-to-Voltage Converter

INTEGRATOR

A simple passive integrator consists of a resistor in series with the signal and a capacitor shunted from the output end of the resistor to common ground, as shown in Figure 7-9(a). A solid stage integrator circuit using a

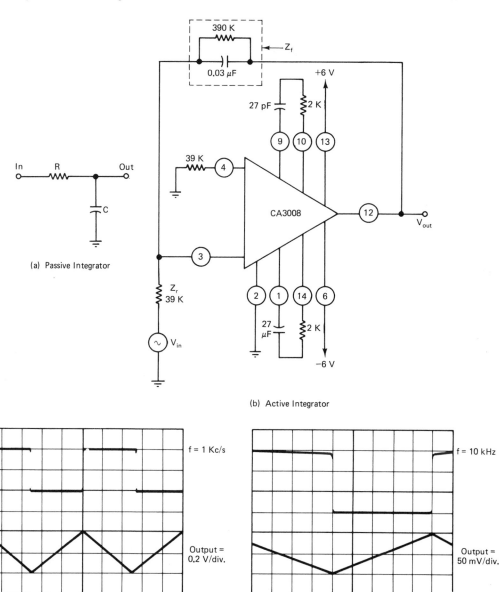

(a) Passive Integrator

(b) Active Integrator

(c) Integrator Waveforms

Figure 7-9. Integrator

CA3008 operational amplifier IC is given in Figure 7-9(b). Feedback from pin 12 to pin 3 through the RC network prevents an offset voltage that cannot continuously charge the feedback capacitor until after the amplifier limits.

The required DC feedback is provided by shunting the feedback capacitor with a resistor to obtain a longer time constant. With the values shown, the gain is limited to 20 dB. As can be see in Figure 7-9(c), the output waveform is triangular and the input waveform is a square wave. (*Courtesy of RCA Solid State.*)

LEVEL DETECTOR

A level detector is a comparator that gives an output when its input is above a certain reference level, and no (or another) output when the input is below a reference. Figure 7-10 shows a simple operational amplifier level detector configuration. The non-inverting input of the op amp is connected to the reference voltage.

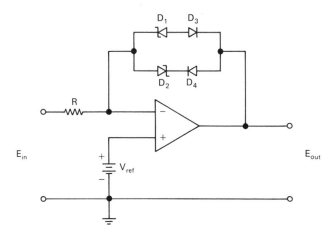

Figure 7-10. Level Detector

If the voltage at E_{IN} is more negative than the reference, the voltage at E_{OUT} is positive. This voltage is limited by Zener diode D2. If E_{IN} is more positive than the reference, E_{OUT} swings negative, but this time the output is limited by Zener diode D1. Diodes D3 and D4 prevent D1 and D2 from conducting during those portions of input when one or the other Zener diode is forward-biased.

If the non-inverting input is shorted directly to ground, with no V_{REF} in the path from the operational amplifier, a *zero-crossing detector* is formed. This circuit yields one output when E_{IN} is above zero volts and another output when the input is below zero.

MICROPHONE PREAMPLIFIER/LIMITER

A Raytheon Raysistor and a 360M op amp are utilized in the microphone preamplifier circuit given in Figure 7-11. The maximum output level is automatically limited to +6 dBm. The Raysistor contains a lamp whose filament is energized by the output signal. As its brilliance varies, so does the resistance of the photosensitive element of the Raysistor. As output level tends to increase, the resistance of the Raysistor across the op amp's feedback path is decreased, reducing its gain. (*Courtesy of Opamp Labs Inc.*)

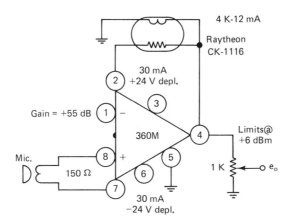

Figure 7-11. Microphone Preamplifier/Limiter

MICROPROCESSOR-RECORDER INTERFACE

The circuit shown in Figure 7-12 permits the recording of information from an 8085 microprocessor onto a tape cassette using a portable cassette tape recorder. The circuit also permits the playing back of the same information, giving a small computer or other MPU-controlled device an inexpensive source of memory.

Data from the 8085 are in the form of tone bursts. These tone bursts are positive in polarity, and would unbalance the capacitive input stage of the recorder if applied directly. So one amplifier (A1) of an LM324 IC is used for isolation, to remove the DC component.

For playback, output is taken from the earphone jack of the recorder, buffered by op amp A2, and inverted by A3. D1 passes the peaks of one of these signals and D2 passes the opposite peaks. The resulting signal is filtered by the RC network and applied to comparator A4, which gives the proper signal level for input to the MPU. (*Courtesy of Intel Corporation.*)

MOSFET ELECTRONIC ATTENUATOR

A dual gate MOSFET can be used as an electronic attenuator when wired into the circuit given in Figure 7-13(a). The input signal is fed to the gate of the MOSFET and the output is taken from the MOSFET drain. The amount of attenuation is controlled with the potentiometer across the 3-volt

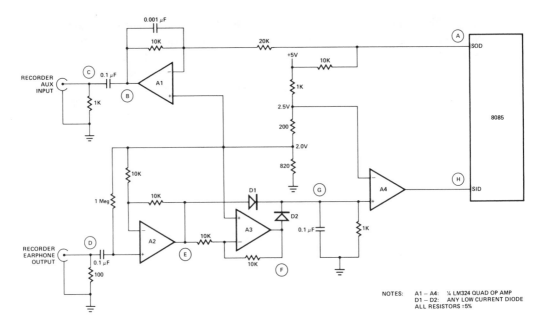

Figure 7-12. Microprocessor-Recorder Interface

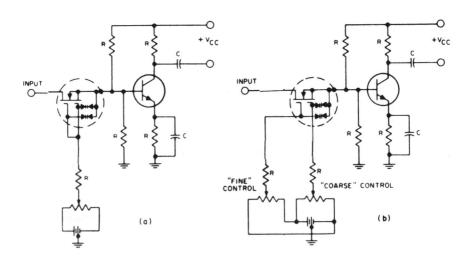

Figure 7-13. MOSFET Electronic Attenuator

center-tapped DC source. This enables adjusting the bias on the two paralleled gates within the −1.5 to +1.5 volt range. The drain-source path through the MOSFET functions as one leg of the voltage divider which includes resistor R from the transistor base to ground.

Two potentiometers are used in the electronic attenuator circuit given in Figure 7-13(b), one for coarse control and the other for fine control. The drain-source resistance of the MOSFET, forms a variable voltage divider. The fine potentiometer is connected to one of the MOSFET gates and the coarse potentiometer is connected to the other gate. (*Courtesy of RCA Solid State.*)

NOISE BLANKER

A noise blanker is included in the better CB transceivers and in most commercial transceivers. A noise blanker minimizes impulse type interference by cutting holes in the signal path. As shown in Figure 7-14, the RF signal at the collector of mixer Q1 is coupled through C1 to a rectifier consisting of CR1 and CR2. When noise pulses are present at the mixer, they are converted into positive pulses which key Q2 into conduction, causing negative-

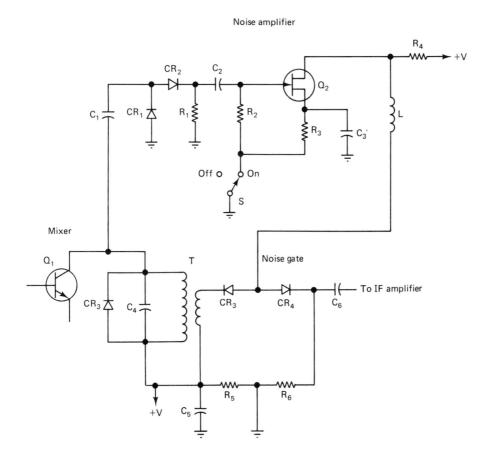

Figure 7-14. Noise Blanker

going pulses to be fed through L to the noise gate. A negative-going pulse at the junction of CR3 and CR4 will reverse-bias these diodes and prevent the IF signal from passing through from the mixer to the IF amplifier. Since these pulses are of very short duration, the IF signal is not usually degraded.

NOISE BLANKER WITH DELAY LINE

A noise blanker chops short-duration holes in the IF signal path of a receiver. The noise is picked up by the receiver antenna and is amplified and then converted into a DC pulse. The problem with a noise blanker is how to get it to know ahead of time when a noise pulse is going to be present and interfere with reception. By the time the noise pulse is converted into a gating pulse, a part of the noise pulse will already have gotten through the IF amplifier. In the circuit given in Figure 7-15, a delay line is inserted between the noise source and the noise gate.

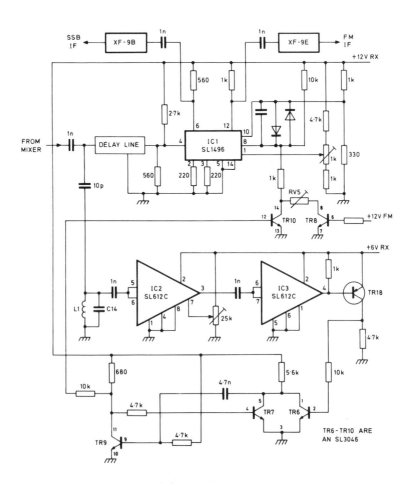

Figure 7-15. Noise Blanker with Delay Line

The IF signal is delayed but the noise pulses that are converted into gating pulses are not delayed. Therefore, the gating pulses can chop holes in the signal path before the noise pulses riding on the IF signal reach the noise gate. The delay line is placed ahead of the crystal filter, which will ring when it receives a noise spike with a fast rise time. But the delay line will have blanked the receiver before then.

In this circuit, the noise gate is contained in the SL1496 IC. The noise is picked up at the output of the receiver mixer and is fed to both the delay line and the input of the two-stage noise amplifier which consists of two SL612C ICs. The amplified noise pulses are buffered by transistor TR18 whose output is fed to a monostable multivibrator (TR6, TR7, TR9) which delivers a 10-microsecond pulse to the noise gate. (*Copyright Plessey Semiconductors.*)

OP AMP ACTIVE NOTCH FILTER

Figure 7-16 illustrates the use of an operational amplifier in an active notch filter. This circuit rejects all signals in a very narrow range, while passing all others, resulting in a sharp "notch" in the frequency response curve. The center frequency of the notch is determined by the formula

$$F = \frac{\sqrt{3}}{RC}.$$

This type of filter is often found in television circuitry, where it is used to sharply attenuate levels at one end of a channel's signal to prevent interference with the next channel. (*Courtesy of Opamp Labs Inc.*)

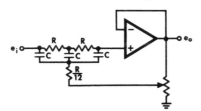

Figure 7-16. Op Amp Active Noise Filter

PULSE SHARPENER

The negative resistance characteristic and rapid switching time of a four-layer diode can be used in the pulse sharpening circuits shown in Figure 7-17. As the input pulse rises, the charge on the capacitor rises until it is great enough to fire the 2N4983 four-layer diode. This creates an output pulse with a very rapid rise or fall time. (*Courtesy of General Electric Company.*)

RF VOLTAGE DIVIDER

The RF voltage divider shown in Figure 7-18 is designed to enable connection of a frequency counter, spectrum analyzer or other device to the output of a radio transmitter. As shown at the left, the voltage divider is built into a PL-250 coaxial plug. The schematic at the right shows that a 10:1 voltage divider is formed when using a 470-ohm resistor and a 47-ohm resistor. (*Courtesy of Hallicrafters.*)

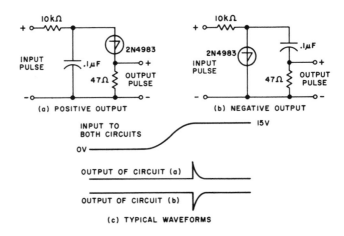

Figure 7-17. Pulse Sharpener

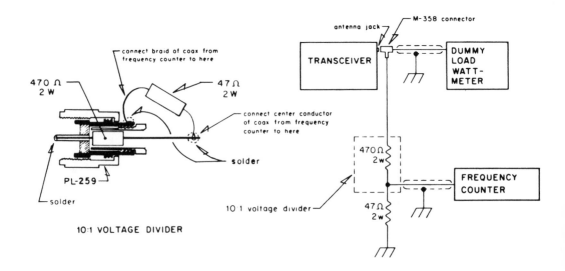

Figure 7-18. RF Voltage Divider

SATURATION LIMITING

When a transistor is driven into saturation, both PN junctions in the device become forward-biased, with the result that a change in the current through one junction no longer affects the current through the other. This can cause clipping of a signal, as in the example of saturation limiting shown in Figure 7-19.

Q1 is an NPN transistor connected as a common-emitter amplifier. The emitter-base junction is forward-biased by V_{BB}, while the collector is reverse-biased by V_{CC}. As the input swings positive, the voltage at the collector falls. Depending on the characteristics of transistor Q1 and its static biasing, the positive-going input signal can actually cause the base to become positive with respect to the negative-going collector. This forward-biases the base-collector junction and causes the output signal to be clipped as shown.

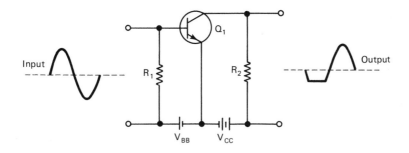

Figure 7-19. Saturation Limiting

SCHMITT TRIGGER

The Schmitt trigger is a comparator circuit. When its input is above a certain level, the Schmitt trigger produces an output; when the input falls below a set level, the output is zero. This feature can be used to convert a sine wave into a square wave, as shown in Figure 7-20.

With no input, Q1 is normally off and Q2 is turned on and saturated. When the input voltage rises to the desired positive level (V_1), Q1 is biased on and the potential at the collector of Q1 drops (becomes more negative). This potential, felt at the base of Q2, causes that transistor to cut off, and the output at the collector of Q2 swings positive. The two transistors remain in this condition, with Q2 turned off and Q1 saturated, until the input falls below level V_2. At this time, Q1 is again cut off, Q2 goes into saturation, and the output falls to its previous level.

In the standby state, when Q1 is off, the Schmitt trigger has high input impedance. However, this impedance falls when the circuit changes state and Q1 is turned on. For this reason, a Schmitt trigger is often preceded by an input buffer stage.

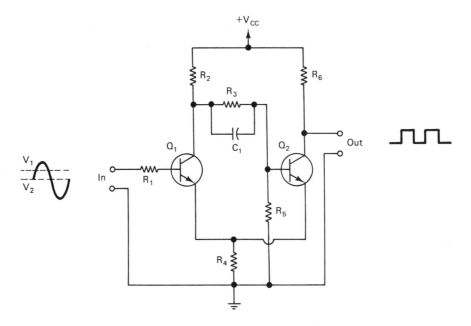

Figure 7-20. Schmitt Trigger

TV TUNER-TO-IF AMPLIFIER INTERFACE

The low-impedance output circuit of a TV receiver tuner can be interfaced with the input of the receiver IF amplifier through the interface circuit given in Figure 7-21. A short length of RG-58A/U 50-ohm coaxial cable is used to take the IF signals from the tuner. These signals pass through a 1,000-pF capacitor, which provides DC blocking, and an 18-ohm resistor to the bridged-T network. (Actually, there are two bridged-T filters; one is tuned

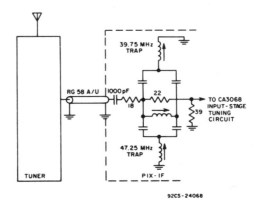

92CS-24068

Figure 7-21. TV Tuner-to-IF Amplifier Interface

to 39.75 MHz and paralleled across the other, which is tuned to 47.25 MHz.) In addition, two wavetraps are used, one tuned to 39.75 MHz and the other tuned to 47.25 MHz. The two wanted signals are the video carrier (45.75 MHz) and audio carrier (41.25 MHz). The traps attenuate 39.75 MHz, the adjacent channel video carrier and 47.25 MHz, the adjacent channel sound carrier frequency, by more than 40 dB. The output of a TV tuner consists of the wanted video IF and audio IF signals, which are 4.5 MHz apart, and also the video carrier of the next lower adjacent channel and the audio carrier of the next higher adjacent channel. When using an outdoor antenna, the presence of these adjacent channel signals generally does no harm because adjacent channels are not allocated in the same area. However, when used with a CATV system, most of which operate on adjacent channels, wavetraps may be required to prevent interference. (*Courtesy of RCA Solid State.*)

VARIABLE ATTENUATORS

Attenuators used to control the volume at speaker locations are of the L-type and the T-type. The L-type maintains a constant load across the source (amplifier output) and the T-type maintains a constant load on both the source and the load (speaker). Attenuators used at speaker locations must handle considerably more power than attenuators used in input lines.

The schematics of four standard variable attenuators are given in Figure 7-22(a). The L-pad is shown at the top of the drawing. The bridged-T is shown just below the L-pad schematic, the straight T-pad circuit is shown below the bridged-T schematic, and the bridged H-pad is shown in the bottom drawing.

It can be seen that the L-pad and the bridged-T use only two variable resistors, while the straight T-pad requires three variable resistors and the

Standard Attenuators

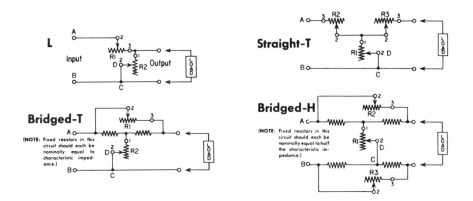

Figure 7-22(a)

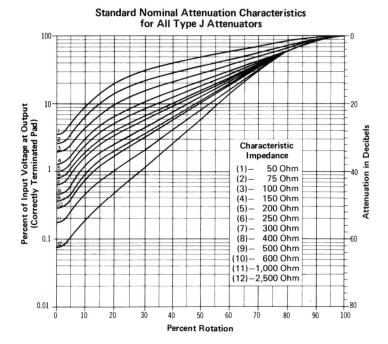

Figure 7-22(b)

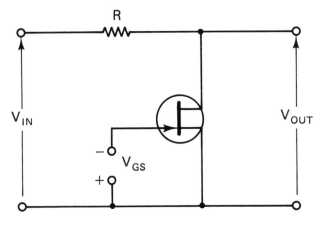

Figure 7-22(c)

bridged-H also requires three variable resistors. The attenuation characteristics of all type-J Allen-Bradley attenuators are shown in Figure 7-22(b). It can be seen that the maximum attenuation of a 50-ohm attenuator is around 24 dB while the maximum attenuation of a 2500-ohm attenuator is greater than 60 dB. Since 600-ohm attenuators are the most widely used type, it is interesting to note that 600-ohm attenuators cause approximately 42 dB at their lowest setting. Bridged-T and L-type attenuators consist of two ganged potentiometers that are rotated together. Attenuators may be inserted at the input and/or the output of an input signal line.

Figure 7-22(c) shows how an FET can be used in a simple attenuator circuit. As long as VCR is operated in the linear portion of its resistance region, varying the amount of V_{GS} bias will vary the attenuation of the circuit. (*Figures 7-22(a) and 7-22(b) courtesy of Allen Bradley;* Figure 7-22(c) © *Siliconix incorporated.*)

CHAPTER 8

Modulators

INTRODUCTION

An unmodulated radio signal can impart only a limited amount of intelligence; it can denote that it exists. But it is modulation that allows such information as voices, music, computer data and teletype signals to be sent over radio waves. Theoretically there could be many kinds of modulation. In practice, however, three kinds are in the widest use: Amplitude modulation, frequency modulation and phase modulation. Other kinds of modulation, such as pulse-code modulation, pulse-width modulation, etc., are also becoming popular in specialized applications such as telephony.

A simple kind of amplitude modulation is produced when a carbon microphone is connected in series with the antenna system of a CW transmitter. The varying resistance within the microphone varies the RF power put out by the transmitter, causing an amplitude-modulated radio signal.

In amplitude modulation, we often speak of the *percent of modulation*. This is a measure of the degree to which the modulating signal changes the carrier. It is proportional to the ratio of the maximum values of the carrier and modulating signal that are present in the final transmitted signal, and is expressed as:

$$M = \frac{\text{Max value of signal}}{\text{Max value of carrier}} \times 100$$

Ideally, modulation is always maintained at 100 percent. When 100 percent modulation exists, the output power rises 50 percent above the unmodulated signal level. Thus, if the transmitter normally delivers a 100-watt signal, output power rises to 150 watts when it is 100 percent modulated. Of this power, the 50 additional watts are contributed by the modulator, which must add 50 watts of audio to the 100 watts of RF. Of the total modulation power, 25 watts are contained in the upper sideband of the signal and 25 watts are contained in the lower.

Figure 8-1 illustrates a signal that is modulated 100 percent, as well as signals that are undermodulated (in this case, 50 percent modulation) and overmodulated (more than 100 percent modulation). When a signal is undermodulated, the power output is reduced and it is not as efficient. When a signal is overmodulated, the modulating signal is distorted.

In the course of broadcasting, the RF signal and the modulating AF signal heterodyne together, producing sum and difference beat frequencies that are also transmitted. For example, suppose a transmitter operates at a center frequency of 1000 kHz, and it is modulated by a signal that goes up to 5 kHz. Regardless of the modulation level, an upper sideband and a lower sideband will be produced in addition to the carrier. The upper sideband extends to 1005 kHz and the lower sideband to 995 kHz.

Since the carrier does nothing except serve as a frequency reference, it is often desirable to eliminate it altogether, as in double sideband (DSB) transmission. And since the two sidebands contain the same information—they are mirror images of each other—higher ef-

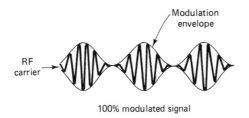

100% modulated signal

Undermodulated carrier signal

Overmodulated carrier signal

Figure 8-1. Amplitude Modulation

ficiency can be obtained by eliminating one sideband as well. To create an SSB (single sideband) signal, the RF carrier and the modulating signal are fed into a balanced modulator. The two signals mix and create an upper sideband (USB) and a lower sideband (LSB), while the carrier is balanced out by the modulator. A filter is then used to eliminate the unwanted sideband.

If the audio modulating signal contains voice frequencies between 300 Hz and 2700 Hz, the single transmitted sideband will occupy only 2400 Hz of bandwidth. Therefore, SSB transmission requires less than half the bandwidth of conventional AM transmission. Since an SSB transmitter emits no signal except when modulation is present, this kind of modulation makes for more efficient use of both power and band space.

In FM, a different modulation scheme is used; instead of varying the amplitude of the carrier signal, the modulator must vary its frequency or its phase, as shown in Figure 8-2. For direct FM, a reactance modulator is used or a varactor diode is used to vary the frequency of the oscillator. In the case of phase modulation, the mod-

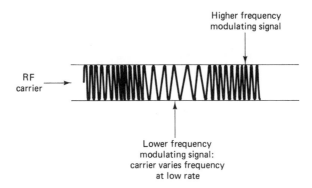

Figure 8-2. Frequency Modulation

ulator is inserted between the oscillator and a frequency multiplier chain. When both the RF and the AF signals are fed in, the output of the modulator varies in phase; this output, when multiplied, appears as an FM signal.

Since in FM the modulating signal produces a changing frequency, the degree of modulation is expressed in terms of frequency deviation. A frequency deviation of 75 kHz is equivalent to amplitude modulation at the 100 percent amplitude level; this deviation means that the total carrier frequency swing will be 150 kHz. Bear in mind that frequency multipliers in FM transmitters multiply both the oscillator frequency and the FM deviation. So if an oscillator operates at 4 MHz and is multiplied 36 times, the output center frequency will be at 144 MHz. And if the frequency deviation at the oscillator or phase modulator is ± 140 Hz, the output signal will then deviate ± 5046 Hz.

This chapter covers various kinds of modulating circuits.

ACTIVE BALANCED MODULATOR

A double sideband suppressed carrier (DSBSC) signal is one that consists of two sidebands (upper and lower). The modulator circuit shown in Figure 8-3(a) uses an MC1496 IC chip to produce a double sideband suppressed carrier output. The gain of the IC can be set by selecting a resistor of the required value for Re, in the degeneration path from pin 3 to pin 2. The carrier is balanced out by adjusting the carrier null potentiometer connected to pins 1 and 4.

The DSBSC signal is similar to a conventional AM signal except that the carrier is suppressed, and all of the intelligence is contained in the sidebands. Figure 8-3(b) shows the relative amplitudes of the carrier and sidebands of the DSBSC signal from this modulator. (*Courtesy of Signetics Corporation.*)

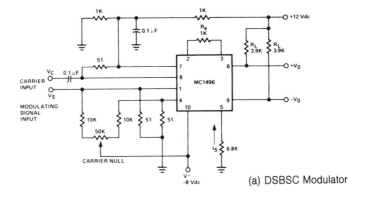

(a) DSBSC Modulator

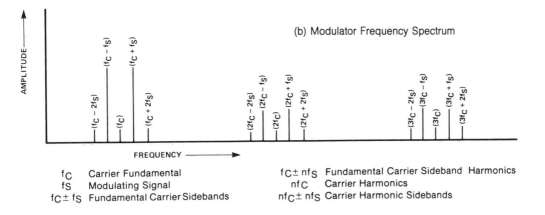

(b) Modulator Frequency Spectrum

f_C Carrier Fundamental	$f_C \pm nf_S$ Fundamental Carrier Sideband Harmonics
f_S Modulating Signal	nf_C Carrier Harmonics
$f_C \pm f_S$ Fundamental Carrier Sidebands	$nf_C \pm nf_S$ Carrier Harmonic Sidebands

Figure 8-3. Active Balanced Modulator

ACTIVE DOUBLE-BALANCED MODULATOR MIXER

The conventional double-balanced modulator utilizes diodes and provides no gain. But a double-balanced modulator that utilizes transistors as shown in Figure 8-4 provides gain and requires a lower level injection frequency signal than the diode type of modulator. In this circuit, a large signal is fed to TR3 and TR4; the signal switches these transistors fully on and fully off. This causes the small signal fed to TR1 and TR2 to mix with the other signal. The output signal consists of the sum and difference frequencies of the two input signals as well as their odd harmonics. This modulator is well suited for use as a receiver mixer.

The output signal of this modulator is DSBSC (double sideband, suppressed carrier). DSBSC modulation is more efficient than conventional AM, which consists of two sidebands (upper and lower) and the carrier. Since the

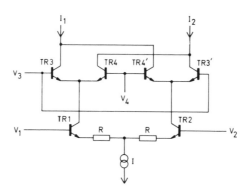

Figure 8-4. Active Double-Balanced Modulator Mixer

carrier conveys no intelligence, it can be eliminated, and the power used for generating the carrier can also be eliminated. This leaves all of the available power in the sidebands, which convey intelligence. DSBSC is not used as widely as SSBSC (single sideband suppressed carrier), which occupies less than half of the bandspace, just like DSBSC. To demodulate a DSBSC or SSBSC signal, it is necessary to generate a signal at the receiver which is at the carrier frequency of the RF signal or at a lower translated frequency (IF). This signal is fed to the detector, where it heterodynes with the translated carrier signal. When the signal is to be at the carrier frequency, it can be applied at the antenna, RF amplifier or mixer.

Since the talk power is contained in the sidebands and the carrier is suppressed, this type of modulator is highly efficient. It will recover the audio from an SSB signal or a DSBSC signal when used as the detector. However, to demodulate SSB and DSBSC signals, a BFO is required to supply an unmodulated signal fed to the double-balanced modulator when used as a detector. Since an SSB signal and a DSBSC signal have their carriers suppressed at the transmitter, the BFO signal is required at the receiver to make the modulation audible. (*Copyright Plessey Semiconductors.*)

AMPLITUDE MODULATOR UTILIZING AN OPERATIONAL TRANSCONDUCTANCE AMPLIFIER

An operational transconductance amplifier is similar to a conventional op amp, but it differs enough to warrant its own classification. The CA-3080 and the CA-3080A are operational transconductance amplifier ICs. The OTA, like a conventional op amp, has a differential input, but it also has an additional control terminal. The output signal is an output current, which is proportional to the voltage difference at the differential input terminals.

A CA3080A operational transconductance amplifier IC is shown in use as an amplitude modulator in Figure 8-5. The carrier frequency signal is fed to the inverting input of the IC. The level of the unmodulated carrier output signal can be established by selection of Rm. When an audio signal is fed to

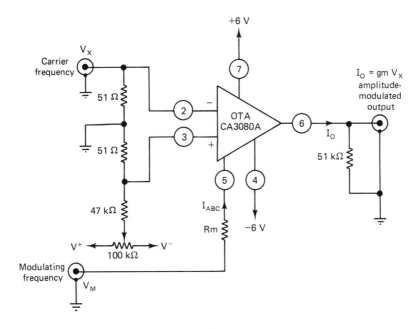

**Figure 8-5. Amplitude Modulator Utilizing
an Operational Transconductance Amplifier**

pin 5 through Rm, the amplifier bias current (ABC) is forced to change. When the AF signal swings positive, the current increases and vice versa. Positive signal excursions increase the transconductance of the amplifier and negative excursions decrease the transconductance.

The characteristic of the operational transconductance amplifier IC is its high output impedance; it is also known as a voltage-to-current-converter. In the modulator application pictured in the diagram, the output is $E_o \times I_L \times G_m$ where E_o is the output voltage, I_L is the load current and G_m is the transconductance in micromhos. (*Courtesy of RCA Solid State.*)

DIODE AMPLITUDE MODULATOR

A single diode is utilized as an amplitude modulator in the circuit given in Figure 8-6. The diode (D), an RF choke coil (L2), the tank circuit (L1-C1) and the modulation input transformer are in series with each other. The RF signal from the oscillator (not shown) is fed in through C2. The audio modulation alternately forward biases and reverse biases the diode, causing the amplitude of the RF signal to vary.

DOUBLE-BALANCED AMPLITUDE MODULATOR

Figure 8-7 shows a double-balanced modulator circuit employing 10 NPN transistors. The RF signal is fed through C1 to the base of Q1, while

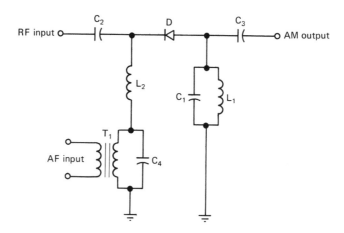

Figure 8-6. Diode Amplitude Modulator

the modulation is coupled through R1 to the base of transistor Q5. The output from the collectors of Q2 and Q4 is a double sideband suppressed carrier signal.

Modulation balance potentiometer R5, wired between the emitters of Q5 and Q6, is adjusted to minimize the modulation signal present in the output, thus holding down distortion. The carrier balance control, potentiometer R7, is adjusted to minimize the presence of the carrier in the output.

DSBSC MODULATOR USING AN LM1596N IC

The circuit given in Figure 8-8 is of a double sideband suppressed carrier modulator employing an LM1596N, a 16-pin dual in-line IC which contains an active balanced modulator circuit. The carrier signal is fed in from an unmodulated oscillator to pin 10 of the IC. The AF modulating signal is fed to pin 1. The output signals at pins 6 and 12 are connected to a balanced circuit tuned to the suppressed carrier frequency. The two signals heterodyne with each other causing an LSB and a USB to be generated. The carrier is balanced out and is not present at the output terminal when the 50,000-ohm balance potentiometer is correctly adjusted. A filter may be used between the output and the next stage to eliminate one of the sidebands and to pass the other sideband to produce an LSB or USB SSB signal. The power supply should deliver +12 volts and −8 volts DC. (*Copyright National Semiconductor Corporation.*)

PHASE MODULATOR

The solid-state version of the Link phase modulator is given in Figure 8-9. It uses an FET instead of a triode tube. The unmodulated RF signal is fed through a capacitor to the FET gate and drain. The AF modulating signal is fed through an RF choke coil to the FET gate. When both the RF signal

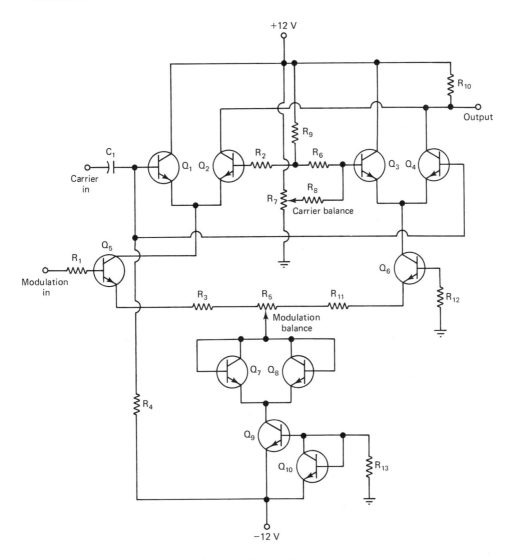

Figure 8-7. Double-Balanced Amplitude Modulator

and the AF signal are present, the RF signal takes two paths, one through the capacitor and one through the FET. This varies the phase of the drain current flowing through the RF choke coil that serves as the drain load. These variations in phase, when fed through a frequency multiplier chain, result in an output signal whose frequency is modulated.

PHASE MODULATOR SYSTEM

Figure 8-10 is a schematic of a phase modulator system whose output signal is taken at J103 and fed to the input of a frequency multiplier chain.

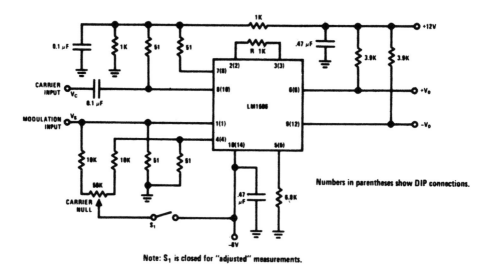

Figure 8-8. DSBSC Modulator Using an LM1596N IC

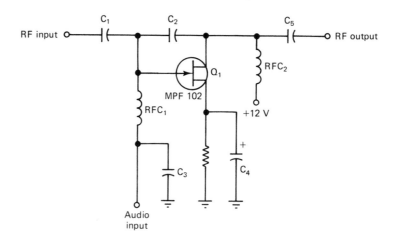

Figure 8-9. Phase Modulator

The microphone signal is fed in at J101 and is coupled through level control R101, C102 and C104 to the base of Q101, which with Q102 forms a ring-tailed pair amplitude limiter and amplifier. The output of Q102 is direct-coupled to emitter-follower Q103 whose output in turn is fed through the twin-diode (CR101—CR102) series limiter to Q104, another emitter-follower whose output is fed to Q105, an active lowpass filter. The output of the active filter is direct-coupled to Q106 whose output is developed across FM deviation control R126. When tone squelch is used, the encoder output is fed to the base of Q107, an emitter-follower whose output is combined with the AF signal present at the wiper of R126.

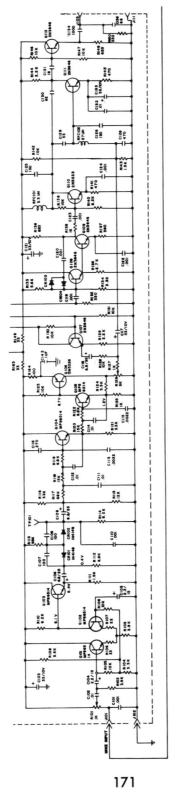

Figure 8-10. Phase Modulator System

171

The signal at the wiper of R126 is coupled through C117 to the audio input of phase modulator Q111. Also fed to the phase modulator is an RF signal through C129 and through C130 to the collector of Q111 and then through C131 to emitter-follower Q112. The RF signal is obtained from the selected TXCO (temperature-compensated crystal oscillator) and is fed through Q108 to the base of Q109. This RF signal is processed through Q109 and Q110, whose output is fed to the phase modulator circuit. (*Courtesy of Aerotron, Inc.*)

RING MODULATOR UTILIZING A DIODE ARRAY IC

The ring modulator circuit given in Figure 8-11 utilizes six diodes of the CA0319 diode array IC although only four are shown in the diagram. (D1 and D2, and D3 and D4, are actually two diodes in parallel.) In this circuit, diodes D1 and D2 conduct during each RF carrier signal half-cycle and diodes D3 and D4 do not conduct. During the other half-cycle D3-D4 conduct and D1-D2 do not. This causes the output amplitude to alternately switch from plus to minus at the frequency of the carrier signal. This results in cancellation of the carrier and generation of upper and lower sidebands when an AF modulating signal is present. (*Courtesy of RCA Solid State.*)

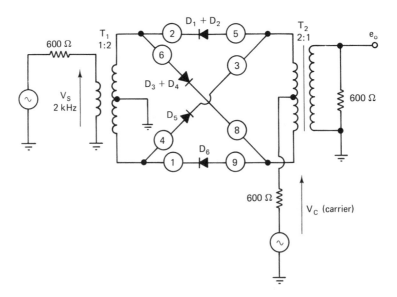

Figure 8-11. Ring Modulator Utilizing a Diode Array IC

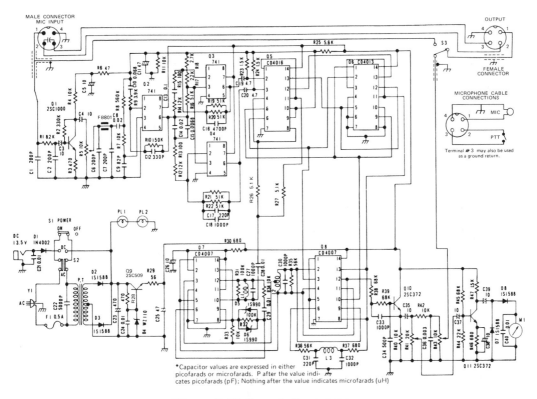

Figure 8-12. Speech Processor

SPEECH PROCESSOR

The Daiwa RF-440 speech processor, whose circuit is given in Figure 8-12, is designed specifically for use with SSB transmitters. It is connected between the microphone output and the transmitter input.

The microphone output is fed in through terminal 1 of the male microphone connector as shown at the upper left corner of the diagram. By setting S1 to the on position, power is applied to the speech processor and the microphone output is fed through R1 and C3 to the base of Q1, an audio amplifier.

The output of Q1 is fed through gain control R5 and a ferrite bead. At the junction of R8 and R7 the output is split and is fed to the differential inputs of Q2, a 741 op amp. Output from Q2 is fed through a pair of 90-degree phase shifters to the inputs of two 741 op amps, whose outputs in turn are fed to the differential inputs of Q5, a CD4016 balanced modulator IC. Also fed to Q5 is a 100-kHz RF signal that is generated by an oscillator associated with Q7, available through Q7. The double sideband suppressed carrier signal is fed to the input of Q7, a CD4007 IC which amplifies and limits the DSBSC signal.

The output of Q7 is fed to the input of Q8, a CD4007 IC which serves as the detector. The 100-kHz RF signal from the oscillator makes it possible to recover the audio signal; this signal is fed to the base of Q10 an audio buffer amplifier. The output of Q10 is fed through output level control R43 and pin 1 of the output jack to the transmitter input. Fed from the wiper of R41 is meter amplifier Q11, whose audio output is rectified by diodes D7 and D8. The rectified audio is fed to meter M1 which has a scale that extends from −20 dB to +3 dB. For optimum performance, the gain and output controls should be set to obtain a 0 dB meter reading.

Briefly, this speech processor converts the voice signal into a double sideband suppressed carrier RF signal which is demodulated so that the recovered audio can be fed to the transmitter. The manufacturer claims a better than 6 dB improvement in talk power. Clipping threshold is less than 2 millivolts at 1000 Hz, frequency response is 300-3000 Hz at 12 dB down, distortion is rated at less than 3 percent at 1000 Hz with 20 dB clipping. Output level can be set to greater than 50 millivolts at 1000 Hz. Power consumption from a 115-volt AC source is approximately 1 watt. When operated from a 13.8-volt DC source, current drain is approximately 55 milliamperes. (*Courtesy of J. W. Miller Division of Bell Industries.*)

SSB MODULATOR

The circuit of the balanced modulator and the speech amplifier used in the Alda SSB transceiver is given in Figure 8-13. The microphone output signal is fed to the base of Q201 which is direct-coupled to Q202, an emitter-follower. The AF signal at the emitter of Q202 is fed through C206 to the cathode of diode D202 and to the anode of D203. The carrier frequency RF signal is fed through potentiometer P201 to the anodes of D202 and D205 through two of the windings of L202, which is an RF transformer wound on a toroid core. The RF signal and the AF signal mix and generate upper and lower sideband signals. The carrier frequency is balanced out when P201 is correctly adjusted. The DSBSC (double sideband suppressed carrier) output signal is fed through a filter (not shown) to a buffer amplifier. The filter blocks passage of the unwanted sideband but allows passage of the wanted sideband. (*Courtesy of Alda.*)

TV RF MODULATOR

An RF modulator like the one whose circuit is given in Figure 8-14 can be used for connecting a TV camera, TV game or a computer data terminal to the antenna terminals of a TV receiver to use as a video monitor. In addition, it can be used to feed the audio output of a microphone (through a preamplifier), phono pickup, scanner monitor or other audio source to enable use of the TV set's audio channel and speaker. The audio signal fed into J1 frequency modulates the 4.5-MHz subcarrier transmitter contained in IC Z1. This heterodynes with the video channel transmitter also contained in IC Z1. The signal fed into J2 amplitude modulates the VHF video transmitter contained in the IC. The RF output can be set to TV channel 3 (61.25/65.75

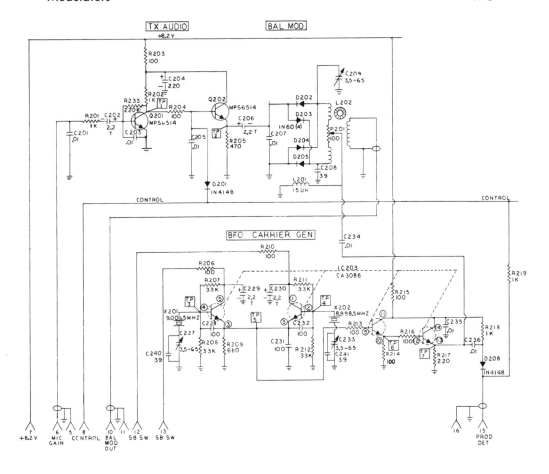

Figure 8-13. SSB Modulator

MHz) or Channel 4 (67.25/71.75 MHz) with switch S2. The device is powered by an integral 9-volt transistor battery. The full 9 volts are fed to pin 3 of the IC and also to the input of voltage regulator IC Z2, which feeds a constant 5 volts to the video level control R2, used to obtain the best quality display. It controls the forward bias on diode CR1 which clips the white peaks in the video signal. Peak RF output is 1500 microvolts, and video input sensitivity is 0.7 to 2 volts peak-to-peak across 75 ohms. FM deviation of the audio channel is 20 kHz per volt of audio input level into the 10,000-ohm input. The RF output is obtained at J3, which can be connected directly to the 75-ohm input of a TV set through a 75-ohm coaxial cable. If the TV set does not have a jack for a 75-ohm coaxial cable, a 75/300-ohm balun must be used to interface the 75-ohm output of the RF modulator with the 300-ohm input of the TV set. (*Courtesy of Radio Shack, a Division of Tandy Corp.*)

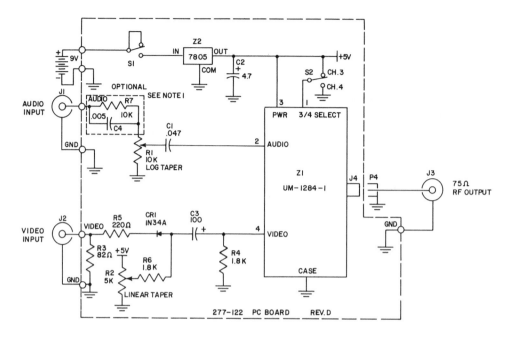

Figure 8-14. TV RF Modulator

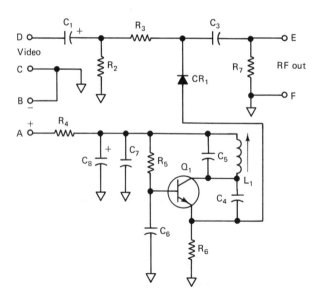

Figure 8-15. Video Modulator

VIDEO MODULATOR

The circuit of a simple video modulator is given in Figure 8-15. This circuit is designed for enabling the use of a TV receiver for viewing the output of a TV camera or other video signal source. Transistor Q1 is the RF oscillator which is tuned to the video carrier frequency of TV channels 3, 4, 5 or 6, depending upon the inductance of L1 and the capacitance of C5. The video signal is heterodyned with the RF carrier signal because of the presence of diode CR1. When C5 is a 15-pF capacitor, the use of a variable inductor for L1 whose inductance can be varied from 0.238 uH to 0.396 uH will enable tuning the oscillator through the 66-85 MHz range. (*Courtesy of Ramsey Electronics, Inc.*)

CHAPTER 9

Demodulators and Detectors

INTRODUCTION

The purpose of a demodulator or detector circuit is to recover the modulation—that is, the audio intelligence, music, or other information—from a radio wave. For this reason, one of the most important characteristics of a demodulator or detector is fidelity. Thus, when such a circuit is designed or selected for use in a particular application, a primary consideration is the amount of distortion or noise that will be allowed. Another consideration is the signal-handling ability of the circuit. And, although the detector is usually placed after several stages of RF or IF amplification, a third consideration is often the sensitivity of the circuit.

Perhaps the easiest way to understand the requirements of an AM demodulator is to study a simple one. Figure 9-1 illustrates a diode used as a very simple AM detector. Input to the circuit is an RF signal that is amplitude-modulated by audio frequencies. The diode conducts during only half of each cycle of RF input, with the result that half the input swing is cut off and half appears across the network formed by C1 and R1.

The time constant of the C1R1 tuned circuit is such that only the audio frequencies are developed across it; capacitor C1, which in a broadcast receiver for the usual AM band is typically about 250 pf, acts as a bypass for RF. The voltage developed across the C1R1 tank, then, varies at the audio rate.

Most commercial and amateur SSB receivers utilize a product detector for recovering the audio from an SSB signal. The output of a BFO (beat frequency oscillator) is combined with the SSB IF signal in the product detector. The BFO is tuned to add the carrier signal that was suppressed at the distant transmitter. (It is not actually tuned to the carrier frequency, but instead is tuned to the translated carrier frequency within the IF passband.) A conventional AM detector may also be used to demodulate SSB signals if a BFO (beat frequency oscillator) is provided.

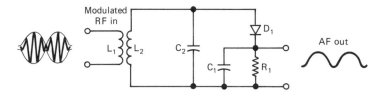

Figure 9-1. AM Detector Principles

The simplest FM detector is the slope detector. This is simply an AM diode detector tuned to one of the slopes of its resonance curve so that the output voltage varies with frequency. Although there are many kinds of FM detector circuits, most modern FM receivers use either the Foster-Seeley discriminator circuit (also called the phase discriminator) or the ratio detector circuit. Since the discriminator circuit is sensitive to amplitude variations, in FM receivers it is usually preceded by a limiter stage.

In the process of frequency-modulation at the transmitter, *pre-emphasis* is intentionally added to higher audio frequencies according to a standard curve. This pre-emphasis is a kind of boosting of certain frequencies; it has been found to result in a better signal-to-noise ratio and better-quality audio reproduction in the receiver at these higher audio frequencies. However, it means that in the FM detector, a network must be added to provide *de-emphasis*, in reverse of the original emphasis. Thus a de-emphasis network, often merely a series-connected resistor followed by a capacitor connected in parallel, is usually added at the output of an FM detector. The time constant of this de-emphasis network is typically around 75 μsec.

This chapter covers various AM and FM demodulators and detectors.

COMBINATION AM AND SSB DETECTOR

Three diodes are used in the combination AM and SSB detector circuit given in Figure 9-2. The IF signal from the IF transformer is fed to the three diodes. One signal path is through D1, the AM detector, and the other path is through D2 and D3 which are connected in series-opposing. When the BFO signal is fed in at the junction of D2 and D3, these diodes will alternately be forward-biased and reverse-biased causing the IF signal path to open and close at the BFO rate. This results in heterodyning of the SSB signal and the BFO signal, making it possible to recover the audio from the SSB signal.

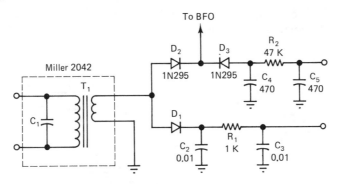

Figure 9-2. Combination AM and SSB Detector

CRYSTAL DISCRIMINATOR

A quartz crystal is used in the FM demodulator circuit given in Figure 9-3. The input is tuned to the center of the IF passband with C1 which is shunted across L. The signal is then split into two directions, one through C2 and diode D1 and a de-emphasis RC network to the audio amplifier (not shown) and the other through C3 and diode D2 to the circuit ground. A variable inductor and the crystal are connected across the signal paths. Any change in the signal frequency results in the generation of a DC output voltage whose polarity depends upon the direction of the FM deviation. When the intercepted FM signal is modulated by a voice signal the output is the receiver audio signal.

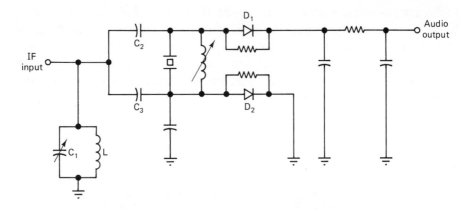

Figure 9-3. Crystal Discriminator

DOUBLE-BALANCED MODULATOR DETECTOR CIRCUITS

Four different detector circuits utilizing active double-balanced modulator ICs are shown in Figure 9-4. In (a) is shown the use of the SL640C IC as an SSB detector. The capacitor connected to pin 5 decouples the sum frequency of the SSB signal and the BFO signal (the recovered audio). The difference frequency is available at pin 6. In (b) is shown the application of

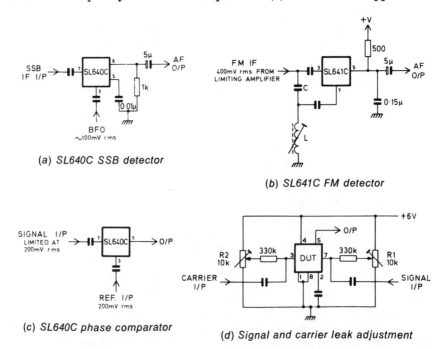

Figure 9-4. Double-Balanced Modulator Detector Circuits

the SL641C IC as an FM detector. In (c) is shown the SL640C in a phase comparator application for demodulating phase modulated signals. And in (d) is shown how the signal and carrier leak can be adjusted. With the carrier signal present, but no intercepted signal present, R1 is adjusted for minimum carrier leak and R2 is adjusted for minimum signal leak. *(Copyright Plessey Semiconductors.)*

DOUBLE-DIODE PRODUCT DETECTOR

In a product detector, the input signal is multiplied together with the signal from a beat frequency oscillator to produce an audio output. Figure 9-5 illustrates a simple product detector using two diodes. The input from the BFO "beats" against the IF, canceling everything but the modulation envelope. Product detectors are often used for detection in SSB receivers because they reduce intermodulation distortion in the AF output, and they do not require a strong input from the BFO.

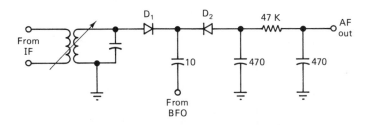

Figure 9-5. Double-Diode Product Detector.

DOUBLE-TUNED FM QUADRATURE DETECTOR

Two quadrature coils are used in the 10.7-MHz FM IF amplifier/demodulator circuit given in Figure 9-6. The circuit uses a CA3089E IC to whose pins 9 and 10 are connected a resonant circuit which is inductively coupled to a second resonant circuit. Both of the windings should have an unloaded Q of 75. Each coil may consist of 20 turns of No. 34 copper wire wound on a 7/32-inch form with E-type tuning slugs. A 0-150 DC microammeter may be connected between pin 13 of the IC through a 330,000-ohm resistor and common ground to serve as a tuning meter. *(Courtesy of RCA Solid State.)*

FM DETECTOR AND AF AMPLIFIER

A single Sprague ULN-2211B IC can be used as the FM demodulator and 2-watt AF amplifier. This 16-pin dual-in-line IC contains the circuitry illustrated in Figure 9-7. Although it is intended for use in TV sets, it can be used in an FM communications receiver following the IF amplifier. The IC requires an input signal of 400 microvolts at the limiting threshold; if the IF is 4.5 MHz, the IF signal can be coupled from the last IF stage of the

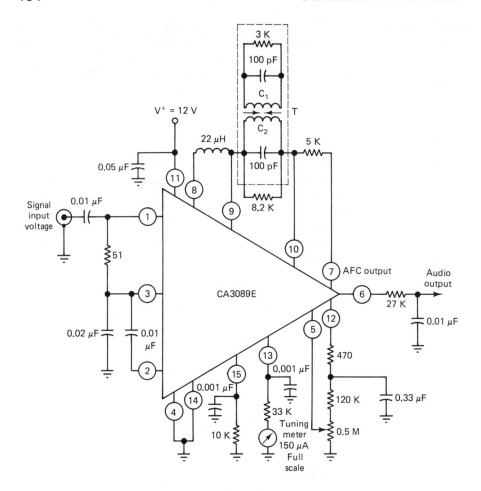

Figure 9-6. Double-Tuned FM Quadrature Detector

receiver through a 4.5-MHz IF transformer to pins 10 and 11 of the IC. If the receiver IF is 10.7-MHz, the same technique can be used, except that a 10.7-MHz IF transformer is required for interfacing the IC input with the receiver IF amplifier output.

The FM IF signal is amplified and limited in amplitude by the IC. The FM signal is then fed to the quadrature detector which recovers the audio from the FM signal. The quadrature network is an L-C tank connected to pins 14 and 15; the coil can be a 10-14 microhenry variable inductor for 4.5 MHz, or a 1-13 microhenry variable inductor for 10.7 MHz. At either frequency, capacitor C has a value of 120 pF. A resistor is not required except when clipping of the FM signal is experienced. The recovered audio at a level of around 300–500 millivolts is fed through an electronic attenuator to the AF amplifier input. Volume level at pin 6 is controlled with a 10,000-ohm potentiometer connected to pin 1 of the IC and the DC supply voltage. *(Courtesy of Sprague Electric Company.)*

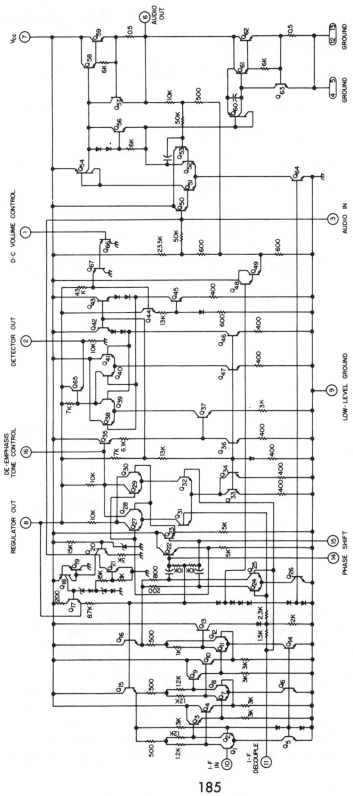

Figure 9-7. FM Detector and AF Amplifier

185

FOSTER-SEELEY DISCRIMINATOR

Figure 9-8 shows a solid-state version of the Foster-Seeley discriminator, or phase discriminator, circuit used as a detector. Because the circuit is sensitive to both AM and FM signals, when used as an FM detector it is usually preceded by a limiter stage.

This circuit consists of a center-tapped transformer and a pair of diodes. Capacitors C3, C4, C5, C6 and C7 have low reactance at the FM intermediate frequency, but high reactance at audio frequencies.

Output from the limiter is developed across the primary winding of discriminator transformer T1 and is also capacitively coupled through C3 to the center tap of the secondary. The voltage developed across the winding connected between terminals 3 and 4 of the transformer is equal to, but 180° out of phase with, the voltage developed across terminals 4 and 5. In addition, the signal across the transformer secondary windings is shifted in phase from the input; the amount and direction of this phase shift are what provide the FM detection of the circuit.

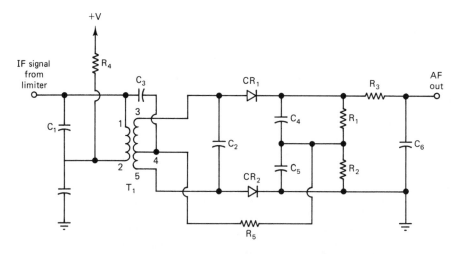

Figure 9-8. Foster-Seeley Discriminator

When the input signal is at the FM IF center frequency of 10.7 MHz, the signals across transformer terminals 3 and 4 and across 4 and 5 are equal in amplitude, 180° out of phase, and also 90° out of phase with the input signal. The rectified voltage developed across R1 and R2 are equal in magnitude, but of opposite polarity with respect to ground.

With a signal above the center IF, the phase angle between the voltage produced by transformer terminals 3 and 4 becomes larger than 90° with respect to the input voltage. At the same time, the phase angle between the voltage produced by terminals 4 and 5 becomes smaller than 90° with respect to the input voltage. This results in the voltage drop across R2 being larger than the voltage drop across R1; and the voltage at the junction of R1 and R3 swings negative with respect to ground.

In a similar fashion, with a signal below the center IF, the phase angle of the voltage produced by terminals 4 and 5 is larger, and the output voltage at the junction of R1 and R3 swings positive. An FM signal, of course, is continually swinging above and below the center frequency, resulting in an audio-frequency output from R3.

LIMITING IF AMPLIFIER/FM DETECTOR

A single Plessey TBA120S IC is used in the limiting IF amplifier and FM detector circuit given in Figure 9-9. This 14-pin dual-in-line IC provides 68 dB of voltage gain and delivers an audio output level of 1.1 volts RMS when demodulating an FM signal with FM deviation of ±50 kHz. Limiting starts when IF input level is 30 microvolts. This IC also contains an electronic attenuator to control the audio output over a 70 dB range through variable resistance connected to pin 5 and common ground. The input signal is fed to pin 14 through a ceramic selectivity filter. (*Copyright Plessey Semiconductors.*)

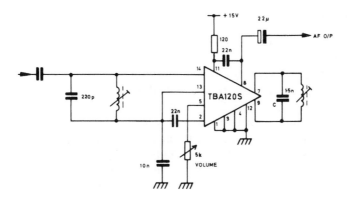

Figure 9-9. Limiting IF Amplifier/FM Detector

QUADRATURE FM DEMODULATOR

Some early FM receivers used the standard Foster-Seeley discriminator circuit as an FM detector. Some used the ratio detector. And, some used an AM detector functioning as a slope detector. In tube-type receivers, the 6BN6 or 12BN6 gated beam tube was used as a highly effective FM detector which had built-in limiting action. Modern receivers use an IC quadrature detector. Figure 9-10 is a diagram of a typical IF system using an IC with an external quadrature coil which is shunted by a fixed capacitor and is connected to terminals 7 and 9 of the ITT TBA 120S IC.

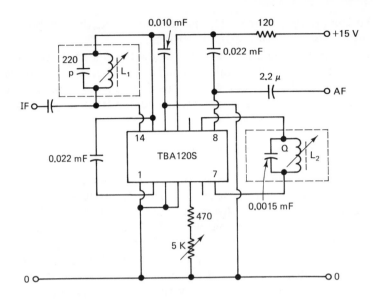

Figure 9-10. Quadrature FM Demodulator

RATIO DETECTOR

The ratio detector is an FM demodulator utilizing two diodes in a circuit that resembles the circuit of the Foster-Seeley discriminator. But it differs in operation and in circuitry from a discriminator. As shown in Figure 9-11, the recovered audio is taken from the center tap of transformer T3 through a tertiary winding. The two diodes are connected in series-aiding and cause a DC voltage to build up across the 5-uF capacitor at the output of the diodes through series resistors. This charge in the capacitor has a limiting effect. It resists changes in signal amplitude, eliminating the need for limiter stages ahead of the detector. The FM signal is demodulated and the recovered audio is fed through an R-C network to the audio amplifier. (*Courtesy of RCA Solid State.*)

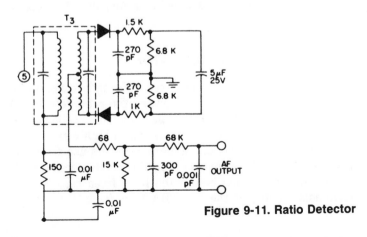

Figure 9-11. Ratio Detector

RING DEMODULATOR FOR SSB SIGNALS

The circuit of a product detector utilizing a four-diode ring demodulator is given in Figure 9-12. The BFO signal is fed through T1 to the balanced input of the ring demodulator. The SSB signal is coupled through T2 and C2 to the ungrounded end of the diamond formed by the diodes. The recovered audio signal is also taken from the same point and is fed through a low-pass filter and C4 to the AF amplifiers (not shown). In this circuit, the BFO signal heterodynes with the SSB signal which makes it possible to recover the audio from an SSB signal whose carrier has been suppressed.

The four diodes are connected in series with regard to the BFO signal. When the signal polarity makes the junction of D1 and D2 positive, D2 and D4 are forward-biased and conduct. When the BFO signal polarity reverses, D1 and D3 conduct. Since the BFO signal is a sine wave, the diodes conduct only after the signal level exceeds the diode barrier voltage. The SSB signal does not have a carrier since it was suppressed at the distant transmitter. The SSB signal is fed through C2 to the junction of D2 and D4. The BFO signal alternately causes the diode to short the BFO signal. This action introduces distortion which is required to cause the two signals to heterodyne with each other.

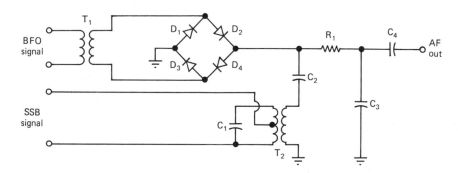

Figure 9-12. Ring Demodulator for SSB Signals

SYNCHRONOUS AM DEMODULATOR

A synchronous detector for reception of DSBSC or SSBSC signals utilizes a BFO signal that is the same as the original carrier that was suppressed at the distant transmitter. In a color TV receiver, the detector stage sees two suppressed carrier sgnals in phase quadrature. Two synchronous detectors employing carriers with a phase difference of 90 degrees enable extraction of the I and Q signals independently from the chrominance signal.

Figure 9-13 illustrates a circuit for a synchronous AM demodulator utilizing an MC 1596K or MC 1496 IC and a ULN 2209 IC. Amplification and limiting of the AM carrier are performed by the ULN 2209 IC, which provides

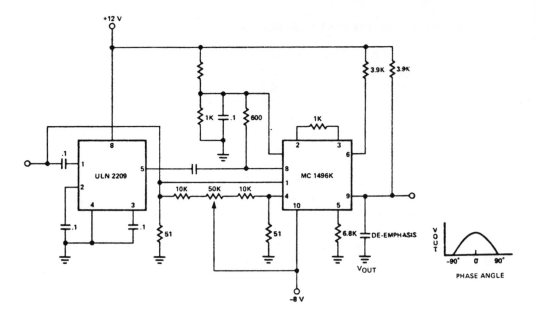

Figure 9-13. Synchronous AM Demodulator

55 dB of gain. The limited carrier is then fed to the MC 1496 IC. This integrated circuit functions as a synchronous detector, delivering the recovered audio signal from pin 9. (*Courtesy of Signetics Corporation.*)

SYNCHRONOUS AM DETECTOR

The AM detector circuit given in Figure 9-14 uses a Plessey SL624C IC in an AM synchronous detector application. This IC contains a limiting amplifier that enables recovering the audio from an AM RF signal with a minimum of noise.

The carrier is separated from the modulation by a limiting amplifier. Then the carrier and sidebands are mixed together to obtain the demodulated output. Nearly all noise products are out of the audio range and therefore are not reproduced. (*Copyright Plessey Semiconductors.*)

TRAVIS FM DEMODULATOR

The Travis FM demodulator circuit given in Figure 9-15 utilizes slope detection for recovering the audio from an FM signal. The IF input signal is split, with one path through C1 to the gate of Q1, an FET. The IF signal is also routed through C2 to the base of Q2, another FET. The FETs provide amplification. The signals are rectified by diodes D1 and D2 in the top leg

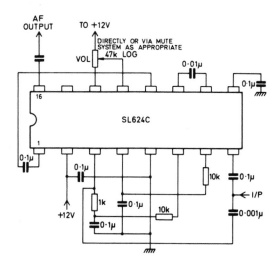

Figure 9-14. Synchronous AM Detector

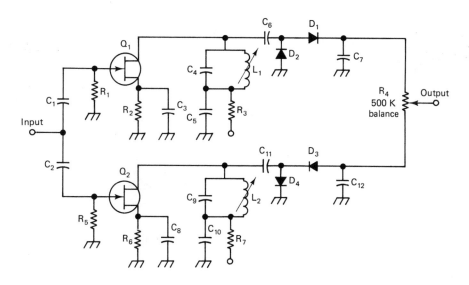

Figure 9-15. Travis FM Demodulator

and by diodes D3 and D4 in the bottom leg. L1 and C4 are tuned to resonate a few kHz above the carrier frequency and L2 is adjusted to resonate a few kHz below the IF carrier frequency. When the IF signal frequency deviates upward, L1 increases the level of the signal and when the frequency deviates downward, L2 increases the level of the signal. Potentiometer R4 is adjusted for minimum distortion and maximum audio recovery.

CHAPTER 10

Electronic Controls

INTRODUCTION

The subject of controls is a large one. In a sense, even the on-off switch on a lamp or radio receiver is a control. However, the controls in this chapter generally cause an electronic device to change state because of a change in time, temperature, pressure, etc.

A control circuit is often a part of "overhead"—that is, it does not contribute directly to the output power, amplification, work, or other primary function of a device. For this reason, control circuitry is usually designed to draw as little current as possible during its quiescent (non-switching) mode. Thus semiconductors, with their relatively low power requirements, are ideal for use in control circuits.

193

In some applications, particularly high-power ones, mechanical relays and stepping motors are used because of their ability to handle high amounts of current. In this case, semiconductors may be employed to control levels or signals, but actual switching is performed mechanically.

However, mechanical switching devices are subject to corrosion, arcing and fatigue. In addition, depending on the application they can themselves draw significant amounts of current. In recent years the power ratings of semiconductor switching devices such as SCRs, diacs and triacs have been increased to the point where 100 percent semiconductor control circuitry is possible in all but the very highest-power applications. And semiconductor devices such as the photon-coupled isolator give a very high degree of isolation between triggering and switching circuits, making these devices ideal for use in computers and other extremely sensitive instruments.

ALARM SYSTEM

Figure 10-1 shows an alarm system comprising a pair of sensor lines connected to a CA 3094 operational transconductance amplifier. In the off or no-alarm state, the potential at terminal 3 is higher than that at terminal 2 of the IC, and terminal 5 is driven with enough current through R6 and the sensor line connected to it to keep the voltage at the output high. If either sensor line is opened or shorted to ground or the other line, the output at pin 6 swings low and activates an alarm. Back-to-back diodes D1 and D2 protect the input terminals of the op amp against excessive voltage swings. (*Courtesy of RCA Solid State.*)

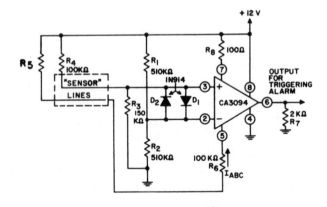

Figure 10-1. Alarm System

AUDIO ALARM

A single VN66AF VMOS transistor is used to directly drive an 8-ohm speaker in the audio alarm circuit given in Figure 10-2. The VMOS output amplifier is fed a 2-kHz tone from an audio generator consisting of two CD4011 gates. The strobe input terminal may be connected to a TTL logic input or other signal source. When operated from a 5-volt DC source, the drain current can rise to as high as 650 milliamperes. (© *Siliconix incorporated.*)

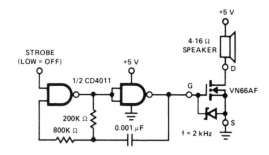

Figure 10-2. Audio Alarm

EMERGENCY LIGHT

In the current shown in Figure 10-3, a GE type C106Y silicon-controlled rectifier wired in the secondary of a transformer is used to trigger emergency lighting in case of power failure. When AC power is applied to the transformer primary, capacitor C1 charges up through resistor R1 and rectifier diode CR1; the negative potential at the gate of the SCR prevents the device from

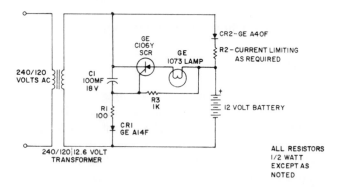

Figure 10-3. Emergency Light

turning on, so the lamp also remains off. However, in the case of a power failure, C1 discharges, allowing the SCR to be triggered on by the battery potential through R3. With the SCR on, battery current flows through the SCR to turn on the emergency light. When AC power is restored, the line voltage biases the SCR and turns it off again. Rectifier CR2 keeps the battery charged. (*Courtesy of General Electric Company.*)

FAIL-SAFE INTRUSION ALARM

An integrated circuit incorporating a photosensitive input is used in the light-controlled intrusion alarm shown in Figure 10-4. The CA3062 has a photo-detector input (pins 9 and 12) and an internal Schmitt trigger. If the light path is broken or the AC line interrupted, the alarm sounds. The battery is kept charged by the $+10V$ power supply while the alarm is being operated from AC. Once triggered, the circuit must be reset with the pushbutton. (*Courtesy of RCA Solid State.*)

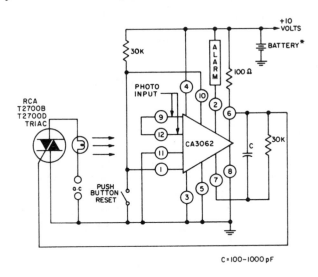

Figure 10-4. Fail-Safe Intrusion Alarm

HOT PLATE CONTROL

Figure 10-5(a) shows the internal block diagram of an SL 443A zero voltage switch, available from Plessey Semiconductors for use as a control in hot plates and hair dryers. The chip contains its own half-wave rectifier to supply internal DC from the AC input at pins 1 and 2.

The period pulse generator produces a single short duration pulse for each completed power cycle. This pulse is used as a clock, and also to switch timing components in the ramp generator, permitting long time constants without using electrolytic capacitors.

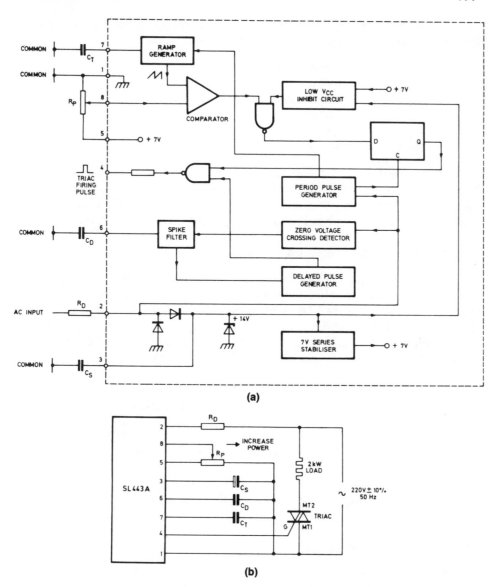

Figure 10-5. Hot Plate Control

The zero voltage crossing detector controls a pulse generator with a delayed output. This delay is necessary because, with inductive or low-power resistive loads, the triac load current may not reach its required holding level at the zero voltage point.

External capacitor C_T determines both delay time and pulse duration, and also filters out spikes in the zero crossing region. The chip will reject spikes having up to 50 percent of the width of a normal triac firing pulse.

Figure 10-5(b) illustrates how the SL443A is used in a hot plate control. Here are component values for the circuit shown:

R_D	8.2kΩ 5% 7W
R_P	100kΩ
C_S	220μF 16V
C_T	0.47μF
C_D	1.5nF ± 10%
TRIAC TAG.225 − 400	

With the value for C_T listed here, the ramp period is about 20 seconds. Through the internal comparator, the setting of potentiometer R_P determines the portion of the ramp period for which the triac is actually conducting, thus controlling the power to the load. (*Copyright Plessey Semiconductors.*)

ISOLATED SOLID-STATE RELAY

A photon-coupled isolator (or opto-isolator) can be used to isolate sensitive TTL computer circuitry from power-carrying circuits in the solid state relay illustrated in Figure 10-6. A 5V pulse through the LED portion of the General Electric 4N40 isolator causes triggering of the SCR in the device. This circuit provides 2500 volts of isolation, and handles 220 VAC loads of up to 10 A. (*Courtesy of General Electric Company.*)

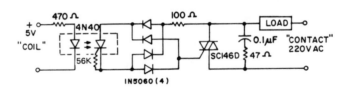

Figure 10-6. Isolated Solid-State Relay

LOW-LEVEL INPUT SENSOR

When signals to a control device are very small and must pass through an environment with high electrical noise, false indications can occur. The circuit shown in Figure 10-7 boosts a low-level input signal by up to 1000 times for application to a control circuit.

A thermocouple strain gauge or transducer is connected in a bridge circuit. Two legs of the bridge are connected to a 10 VDC source, while the other two are wired directly through a twisted pair cable (to minimize noise) into the high and low inputs of a Datel data acquisition integrated circuit. This IC contains a differential amplifier that rejects common-mode noise and amplifies only the difference between the two input signals. (*Courtesy of Datel-Intersil.*)

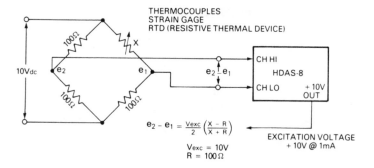

Figure 10-7. Low-Level Input Sensor

NIXIE TUBE DRIVER

A high-voltage NPN transistor is used to drive each segment of a Nixie indicating tube in the circuit shown in Figure 10-8. A positive input to the base of a transistor turns the transistor on, and collector current flows through the desired tube element. (*Courtesy of General Electric Company.*)

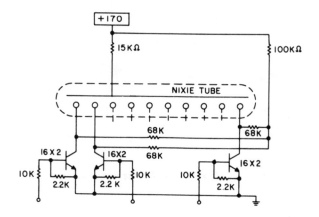

Figure 10-8. Nixie Tube Driver

PHOTO DRIVER FOR A TRIAC

A Motorola MRD 3010 or 3011 optically triggered triac driver may be used to control a triac with a resistive load, as shown in Figure 10-9. When the triac is used to control an inductive load, the R-C filter should be added. In these drawings the device at the left represents the driver; the triac seen in the load circuit is not a part of the trigger device. When sufficient light reaches the trigger, it will conduct and apply a trigger voltage to the triac. (*Courtesy of Motorola, Inc.*)

RESISTIVE LOAD

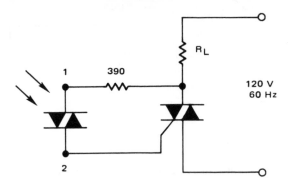

INDUCTIVE LOAD

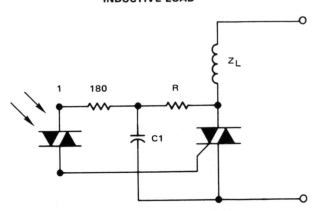

Figure 10-9. Photo Driver for a Triac

PIN DIODE ANTENNA SWITCH

A pair of PIN diodes can be used in the circuit given in Figure 10-10 to switch an antenna from a receiver to a transmitter. When using a Unitrode UM9401 PIN diode as D1 and a Unitrode IN5757 as D2, the receiver isolation will be greater than 40 dB at 50 MHz and insertion loss will be 0.1 dB when the bias current flowing through D2 is 50 milliamperes. This antenna switch can be used with transmitters delivering up to 50 watts of output. If R is a 240-ohm, 1-watt resistor, application of 12 volts DC to bias 2 will cause diode D2 to switch on and route the antenna signal to the receiver. When 12 volts DC is applied to bias 1, D1 will switch on and route the transmitter signal to the antenna. (*Courtesy of Unitrode Corporation.*)

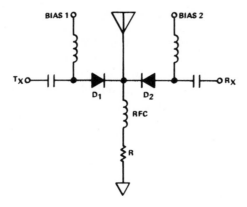

Figure 10-10. PIN Diode Antenna Switch

SOLID-STATE RELAY

A solid-state relay contains no moving parts and has no contacts that require cleaning. The functional diagram of a typical solid state relay is given in Figure 10-11. The DC control voltage (4 volts and higher) is fed to terminals 3 and 4. The control current flows through a GaAs LED connected in series with a 200-ohm current limiting resistor. The light given off by the LED is intercepted by a photo detector IC, causing it to develop an enable signal. A zero crossing detector generates trigger pulses that are used to trigger the two back-to-back connected SCRs. Once triggered, each SCR continues to conduct load current until the AC power line cycle reverses the polarity of the voltage across the SCR, causing it to stop conducting. The SCRs control a triac which controls the load.

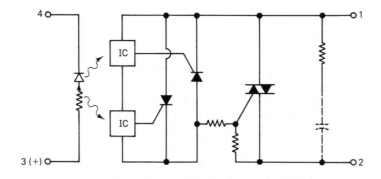

Figure 10-11. Solid-State Relay

THERMISTOR TEMPERATURE CONTROL

In the temperature regulator and controller shown in Figure 10-12, a linear thermistor network (LTN) is used to control an electric heater. The desired temperature is selected by the voltage picked off at the wiper arm of R3 and applied to the noninverting input of the op amp. As the temperature falls, the resistance of the thermistors in the LTN rises; this causes the output of the op amp to swing positive, turning on Q1 and Q2 and the heater. The range of the circuit shown is −5°C to 45°C. (*Courtesy of Fenwal Electronics.*)

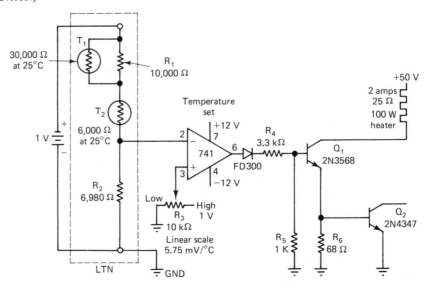

Figure 10-12. Thermistor Temperature Control

TIME DELAY CIRCUIT

A pair of unijunction transistors and an SCR are connected in the time delay circuit shown in Figure 10-13. When power is applied to the circuit, the SCR is turned off and no path is immediately available from the load to ground.

Capacitor C1 and resistor R1 determine the delay time of the circuit, and can be selected for a delay of from 0.3 milliseconds to over three minutes. After power is applied, C1 charges through R1 until the voltage at the emitter of the 2N494C is sufficient to fire the UJT. The pulse at the lower base of the UJT then fires the SCR, and the supply voltage (minus about one volt) is placed across the load. The 2N491 unijunction transistor is used as a relaxation oscillator which pulses the upper base of the 2N294C, thus effectively reducing the minimum firing current necessary for the 2N294C and providing a greater range of delay time. (*Courtesy of General Electric Company.*)

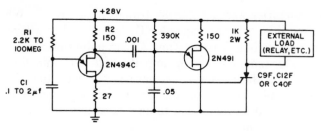

PRECISION SOLID STATE TIME DELAY CIRCUIT

Figure 10-13. Time Delay Circuit

TIMER WITH METER READOUT

Figure 10-14 shows the circuit for a presettable timer with a meter that indicates elapsed time. The circuit uses an RCA CA 3094 operational trans-

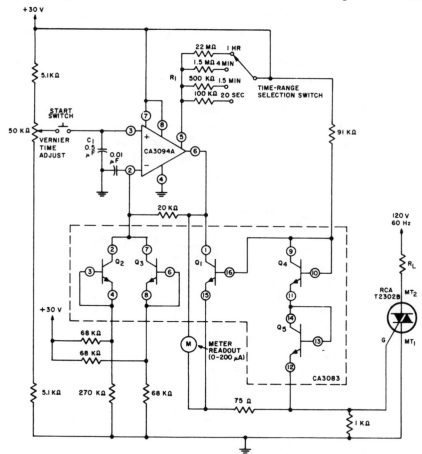

Figure 10-14. Timer with Meter Readout

conductance amplifier and a CA 3083 transistor array IC. When the start switch is closed momentarily, capacitor C1 is charged to a level determined by the setting of the 50K Vernier Time Adjust potentiometer. C1 discharges slowly through terminal 3 of the op amp at a rate determined by the resistor selected for R1. As the capacitor discharges, the collector of transistor Q1 and the output at pin 6 of the op amp are high. Diode-connected transistors Q4 and Q5 are wired so that Q1 acts as a constant-current source to gate the triac on at this time.

As Q1 discharges, the output at pin 6 of the op amp gradually falls, causing the meter to indicate less current flow. When the output is no longer sufficient to hold Q1 in saturation, the transistor turns off, the gate current to the triac is interrupted, and the timer output switches off. (*Courtesy of RCA Solid State.*)

TOUCH-CONTROLLED SELECTOR

The circuit of a touch-controlled selector is given in Figure 10-15. This circuit is designed specifically for use as a TV or FM receiver channel selector. It uses a 26-pin dual-in-line Plessey ML238 IC. It can be used for selecting up to 8 channels and utilizes 8 LEDs as indicators. As shown in the diagram, a diode and a current-limiting 1000-ohm resistor are used in series with each LED. There are 8 touch plates; when a finger is placed on one of them, the drop in resistance across the touch-plate triggers the selected channel. At each channel position, a potentiometer is connected which enables setting the varactor voltage to tune the receiver to the selected channel. The system draws up to 6 milliamperes when connected to a 34-volt DC power source. (*Copyright Plessey Semiconductors.*)

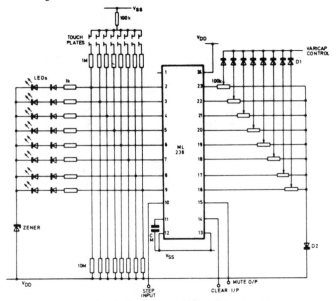

Figure 10-15. Touch-Controlled Selector

TOUCH-SENSITIVE SWITCH

Figure 10-16 shows a diagram of a simple touch-sensitive switch for use in controls. When power is applied to the circuit, FET T1 is biased on and conducts charging the 0.47 μF capacitor and turning off T2.

Touching the sensor provides enough capacitance to turn on T2, discharging the 0.47 μF capacitor through the lower 4.7 megohm resistor. T2 is held on by the potential felt through the 4.7 megohm resistor between the drain of T2 to the gate of T1.

Touching the sensor once more causes the positive potential on the 0.47 μF capacitor to be felt at the gate of T1, turning it on again and turning T2 off. (*Copyright © International Telephone and Telegraph Corporation.*)

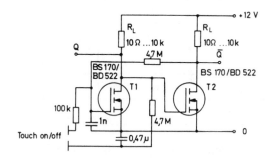

Figure 10-16. Touch-Sensitive Switch

UNIVERSAL MOTOR SPEED CONTROL

Figure 10-17 shows a General Electric type C106 SCR as the heart of a speed control for small universal motors (such as are found in mixers, sewing machines, etc.). The RC network made up of capacitor C1 and resistors R1 and R2 determines a reference voltage for triggering the SCR. This reference is balanced against the counter electromotive force set up by the motor.

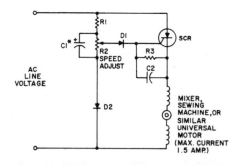

Figure 10-17. Universal Motor Speed Control

As the motor speed slows down because of heavy loading, the counter emf falls, and the reference ramp triggers the SCR earlier in the AC cycle. This applies more voltage to the motor, causing its speed to increase again.

Here are some representative circuit values:

Line Voltage	120V	240V
R_1	47K	100K
R_2	10K	20K
R_3	1K	1K
C_1	1μF, 50V	1μF, 100V
C_2	0.1μF, 50V	0.1μF, 50V
D_1	1N5059	1N5060
D_2	1N5059	1N5060
SCR	C106B1	C106D1

The maximum current capability of this control is 1.5 A. (*Courtesy of General Electric Company.*)

VALVE OR SOLENOID CONTROL

Many commercially available valves, motor actuators, and solenoids require potentials in the 24-30 volt range for actuation, at currents that can be well over an ampere. Figure 10-18 shows a circuit that can be used to drive these controls.

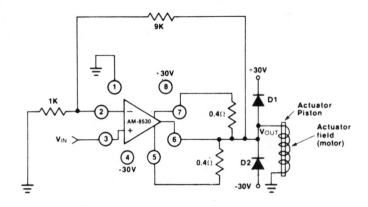

Figure 10-18. Valve or Solenoid Control

The heart of the circuit is a Datrel AM-8530 integrated circuit, an IC in which the output from an operational amplifier is fed into a power amplifier. With an input to V_{IN} of $+2.4$ volts, V_{OUT} is 24 volts; to reverse the direction of piston or solenoid travel, the input should be negative. The two 0.4-ohm resistors limit current to protect against shorts, while diodes D1 and D2 absorb the inductive kick of the motor or solenoid during turn-on and turn-off. (*Courtesy of Datel-Intersil.*)

CHAPTER 11

Digital Logic and Computer Circuits

INTRODUCTION

Just as ancient Oriental philosophers divided the world into yin and yang, sun and moon, modern electronics may be thought of as divided between analog and digital circuitry. Analog signals can consist of many different levels and are continuous; most of the circuits in the other chapters of this book are analog. Digital circuits, however, are quite different.

Digital and logic circuits, which are found not only in computers and calculators, but in control applications, communications, and other uses as well, are characteristic for their ability to detect and respond to the difference between only two states. The actual states used can vary widely: a relay may be open or closed, a circuit can be "on" or "off," or a voltage can be "up," (high) or "down" (low). Some circuitry uses what is sometimes called "positive logic," where a positive voltage is the on state, while a zero or less positive voltage is the off state. Other circuitry employs "negative logic," where a greater *negative* potential is the on (or high) state, while the off state is represented by a less negative potential.

Although the individual circuitry is comparatively simple, even the smallest of computers uses thousands of digital circuits, or "gates," interconnected in patterns which can perform such tasks as addition, subtraction, and memory. Although the operation of each individual gate might, by itself, seem trivial, the speed of operation of these gates, together with their tremendous quantity, makes a computer able to process a great volume of information quickly.

In a digital computer, a single piece of information is represented by the on or off, or the up or down state. This single piece of information is known as a *bit*. Many of these bits together make up a word or byte. The size of a word depends on the computer; in a microcomputer or so-called personal computer, the size of a word is usually 8, or sometimes 16, bits long.

When ENIAC, the world's first all-electronic digital computer, went into operation in the late 1940's, it was packed with thousands of mechanical relays and vacuum tubes, with the result that the computer was as large as a house and as noisy as a factory. Its 18,000 tubes gave off an eerie red glow as well as a considerable amount of heat. In addition, ENIAC's program memory size was relatively small by today's standards—16,000 bits, about the same size as the memory of some modern hand-held programmable calculators—and very slow.

It is the use of semiconductors that has made possible the small, powerful computers of today. Initially banks of transistors, diodes, and other components were actually wired into gates, and these gates in turn were attached together in the desired configuration. More recently, digital logic circuitry has been placed on integrated circuits, or "chips," which consist of layers of semiconductor material through which channels have been left during manufacture. The channels form the individual transistors and gates in the semiconductor material, and can even form the desired connections among the gates.

When 10 gates or less are placed on a single chip, it is called SSI, or small-scale integration. MSI, medium-scale integration, has up to 100 gates on a single chip. An LSI (large-scale integration) chip can have 1,000 or more separate circuits packed onto a single wafer smaller than a fingernail. And VLSI (very large-scale integration) can put 10,000 or more circuits on a chip.

There are many different "families" of digital logic. These are usually described by their main elements, such as DTL (diode-transistor logic) or CMOS (complementary metal-oxide semiconductor). A designer selects a logic family based on cost, reliability, and parameters such as speed, noise immunity, and fanout (the number of logic elements that can be driven by the output from a single gate). In addition, the designer may need to consider whether one logic family can be interfaced with another—that is, whether their signals and power sources are compatible.

Digital Logic Terminology

The operation of many types of digital circuits can be described by means of a *logic symbol*, a *logical equation*, or a *truth table*. In a logic symbol, the actual connections of the gate are shown, as seen in the AND gate illustrated in Figure 11-1. In this gate, when the input to A *and* the input to B are both logical "ones," the output at C is also a "one." When the input to *either* A or B is a "zero," however, the output at C will also be a "zero."

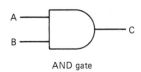

AND gate

Figure 11-1. AND Gate

A small circle on an input or output line next to the gate indicates that the signal there is inverted. So in the diagram of the NAND (Not AND) gate shown in Figure 11-2, when the inputs to both A *and* B are logical ones, the output at C is inverted to a logical zero. If the input at either A or B is a logical zero, however, the output at C is a logical one.

The logical equation for the AND gate in Figure 11-1 is $A \cdot B = C$. This means "A *and* B yield output C." A dot ($\cdot$) in a logical equation is the symbol for an AND function; a plus sign ($+$) is the symbol for an OR function. A straight line over the symbol for an input or output is read as "Not," indicating that input or output is inverted. So the logical equation $A + \overline{B} = C$ is read as "A *or* Not B (that is, when B is a logical zero) yields C."

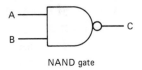

NAND gate

Figure 11-2. NAND Gate

A truth table for a gate illustrates every possible combination of inputs to that gate, along with the resulting outputs. A truth table for the NAND gate in Figure 11-2 is shown in Figure 11-3.

A	B	C
0	0	1
1	0	1
0	1	1
1	1	0

NAND gate truth table

Figure 11-3. NAND Gate Truth Table

A-TO-D CONVERTER

An A-to-D converter changes an analog signal, such as a changing voltage, into a digital signal. A-to-D converters are valuable interfaces between analog and digital circuitry, and they can, with the proper offset and scaling circuits, digitize a wide variety of analog signals for application to a computer or microprocessor.

The 8-bit A-to-D converter shown in Figure 11-4 uses a ZN425E D-to-A converter chip coupled to a ZN424P operational amplifier, which is used as a voltage converter. On the negative edge of the convert command pulse, the counter is set to zero and the status latch to logical one. On the positive edge, the gate is opened, enabling clock pulses to be fed to the counter input at pin 4 of the ZN425.

The analog output at pin 14 of the ZN425 climbs until it equals the voltage on pin 4 of the op amp. At this point, the comparator output goes low and resets the status latch to inhibit further clock pulses. This logical

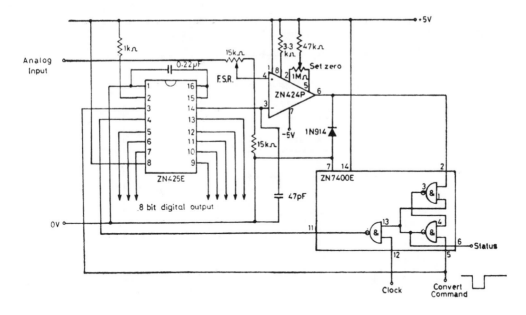

Figure 11-4. A-to-D Converter

zero from the comparator indicates that the 8-bit digital output produced by the ZN425E is now a valid representation of the analog input voltage.

The 47pF capacitor at pin 3 of the comparator stops any positive-going "glitches" from prematurely resetting the status latch. This capacitance is in parallel with the ZN425 output capacitance; together they form a time constant with the ZN425 output resistance, and this time constant is the main limit to the maximum clock frequency. Using the ZN424 as the comparator, the clock frequency can run as high as 100 kHz. The A-to-D conversion time varies with the input, and is a maximum for full scale input:

$$\text{Maximum conversion time} = \frac{256}{\text{Clock frequency in Hz}}$$

(Courtesy of Ferranti Semiconductors.)

CMOS FLIP-FLOP

The CMOS RS (reset-set) flip-flop in Figure 11-5 uses pairs of complementary N-channel and P-channel metal-oxide semiconductor field-effect transistors (MOSFETs) to form a memory unit that consumes almost no power.

When a logical one (+ potential) appears on the SET line, N-channel transistor Q2 is gated on, while P-channel transistor Q1 is gated off. Since the source-drain path of Q2 now offers little resistance, the $\overline{Q}$ line, at the drain of Q2, is effectively at ground (a logical zero). The zero on the $\overline{Q}$ line

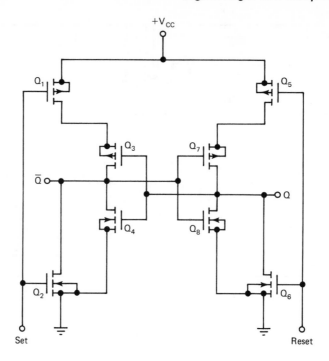

Figure 11-5. CMOS Flip-Flop

cuts off Q8 and turns on Q7. Since at the same time Q5 is on and Q6 is off because of a logical zero on the RESET line, Q is effectively close to + V_{CC}—that is, a logical one.

Even if the SET pulse disappears, the flip-flop maintains its same state. The Q line is held at a logical one through the series drain circuits of Q5 and Q7, and $\overline{Q}$ is latched at a logical zero through Q2 and Q4. The flip-flop remains in this state until a RESET pulse appears.

A logical one (+ potential) on the RESET line turns off Q5 and turns on Q6, and Q goes to a logical zero through the drain circuit of Q6. This causes Q3 to turn on, and $\overline{Q}$ approaches + V_{CC} (logical one) through the drain circuits of Q1 and Q3.

Notice that there is no appreciable current flow in this flip-flop. Since measurable current drain is only a few picoamperes, CMOS circuitry is often used to provide "continuous memory" in calculators and other battery-powered devices. Even with the power switch off, a slight trickle of battery current, enough to maintain the information in CMOS flip-flops, is permitted to flow.

CMOS INVERTER

Figure 11-6 illustrates the operation of a CMOS (complementary metal oxide semiconductor) inverter. The circuit uses both a P-channel field-effect

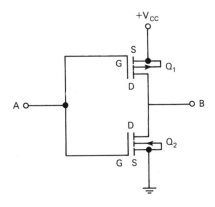

Figure 11-6. CMOS Inverter

transistor (Q1) and an N-channel FET (Q2). With a low (ground) input, the voltage at the gate of Q2 is insufficient to turn that transistor on. At the same time the P-channel FET (Q1) is forward-biased, and the drain terminal of Q1 is near $+ V_{cc}$, yielding a logical one (high) output at B.

When the input at A is high (+), Q1 is cut off as its gate becomes less negative. Q2 is forward-biased by the positive potential at its gate, and the drain voltage of Q2 falls to near ground. Thus, with a one input at A, the output at B falls to a logical zero.

Since the pair of complementary transistors has been connected in series, the circuit draws virtually no drain current. The only power dissipated is during the switching from state to state. Although CMOS has higher manufacturing costs and lower packing density than either PMOS or NMOS, it is very effective in battery-operated applications and other uses where the power source is limited.

CSDL NAND GATE

The NAND gate shown in Figure 11-7 uses Current Switching Diode Logic (CSDL) to make the gate sensitive to small levels of input change. This sensitivity is possible because transistor Q1 is isolated from inputs A, B, and C.

With a low signal (logical zero) to any input, current flows through the diode (D1, D2 or D3) and R1, keeping diode D4 cut off. This in turn prevents the base current of Q1 from rising high enough to forward-bias the transistor. With Q1 cut off, the level at its collector and at the output, D, is a high voltage—a logical one.

When the inputs to A *and* B *and* C are all positive (logical ones), the input diodes D1, D2 and D3 are all cut off. The potential at the cathode of D4 rises, D4 conducts, and enough base current now flows through the emitter-base junction of Q1 to forward-bias the transistor. This results in

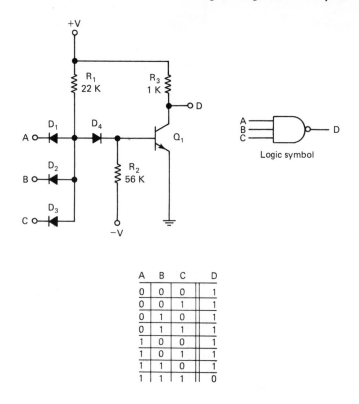

A	B	C	D
0	0	0	1
0	0	1	1
0	1	0	1
0	1	1	1
1	0	0	1
1	0	1	1
1	1	0	1
1	1	1	0

Figure 11-7. CSDL NAND Gate

high collector current and low collector voltage, and the output at D falls to a logical zero. Thus:

$$A \cdot B \cdot C = \overline{D}.$$

DIODE LOGIC AND GATE

Diode logic uses a combination of diodes, usually silicon, as circuit elements for logic gates. In the simple diode logic AND gate shown in Figure 11-8, the inputs to A *and* B *and* C must be logical ones (+ voltage) for the output at D to be a one. If any of the inputs falls to a logical zero, the diode associated with that input will conduct, and the output at D will fall to a zero. Thus:

$$A \cdot B \cdot C = D.$$

DIODE LOGIC OR GATE

The OR gate shown in Figure 11-9 also uses diode logic. When a logical one (positive voltage) is applied to input A *or* B *or* C, the diode associated with that input conducts, causing the voltage to appear at the output, D. Thus:

$$A + B + C = D.$$

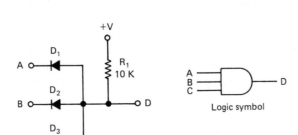

A	B	C	D
0	0	0	0
1	0	0	0
0	1	0	0
0	0	1	0
1	1	0	0
1	0	1	0
0	1	1	0
1	1	1	1

Figure 11-8. Diode Logic AND Gate

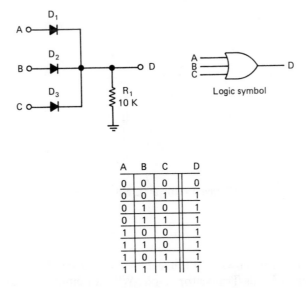

A	B	C	D
0	0	0	0
0	0	1	1
0	1	0	1
0	1	1	1
1	0	0	1
1	1	0	1
1	0	1	1
1	1	1	1

Figure 11-9. Diode Logic OR Gate

DTL NAND GATE

Diode-Transistor Logic, or DTL, employs diodes and transistors as elements of logic gates. In the DTL NAND gate shown in Figure 11-10, when the input to any of diodes D1, D2 or D3 is low (a logical zero), there is insufficient forward bias at the base of NPN transistor Q1, and the transistor is cut off. Since Q1 is cut off, there is no collector current and high collector voltage, and the output at D is high (a logical one).

When the inputs to A *and* B *and* C all become high (logical ones), the voltage at the base of Q1 rises, forward-biasing the transistor. Q1's collector current rises, and the voltage at the collector and at output D falls to a logical zero. Thus:

$$A \cdot B \cdot C = \overline{D}.$$

DTL is one of the oldest types of logic families. It has gate propagation delay of about 30 ns and limited fanout, but most versions of it are compatible with TTL and it is widely used.

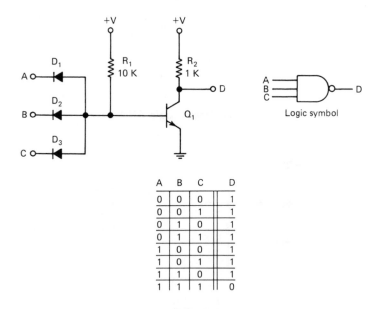

A	B	C	D
0	0	0	1
0	0	1	1
0	1	0	1
0	1	1	1
1	0	0	1
1	0	1	1
1	1	0	1
1	1	1	0

Figure 11-10. DTL NAND Gate

DTL NOR GATE

The NOR gate shown in Figure 11-11 uses a combination of semiconductor devices in Diode-Transistor Logic. With all inputs at a low logic state (logical zero), transistor Q1 is cut off and the output at D is a high positive

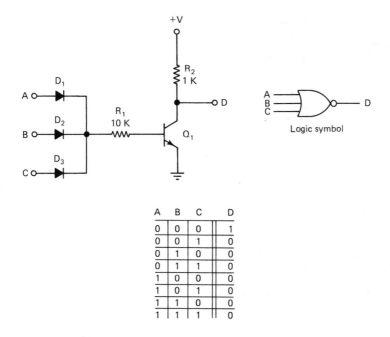

A	B	C	D
0	0	0	1
0	0	1	0
0	1	0	0
0	1	1	0
1	0	0	0
1	0	1	0
1	1	0	0
1	1	1	0

Figure 11-11. DTL NOR Gate

voltage (logical one). If a positive voltage in the form of a logical one is applied to input A *or* B *or* C, the diode (D1, D2 or D3) associated with that input conducts, giving a path for current through the emitter-base junction of Q1. This forward-biases Q1, causing a rise in collector current and a drop in collector voltage at output D to a logical zero. Thus:

$$A + B + C = \overline{D}.$$

D-TO-A CONVERTER

A D-to-A converter is a circuit which changes a digital input to an analog output—that is, the output voltage varies with the number of input bits that are set (true). D-to-A converters are widely used in control circuits and other applications involving computers and microprocessors.

Figure 11-12 shows a basic D-to-A converter. The integrated circuit is a ZN428 chip, which accepts a 8-bit input on pins 11-16 and 1-2. The most significant bit (MSB) is on pin 11 and the least significant bit (LSB) is applied to pin 2. Pin 8 is connected to analog ground (A), while pin 9 is connected to digital ground (D). Input bits can be accepted only when the ENABLE line, pin 4, is low.

With the ENABLE line set to a logical zero, as the input bits change, the voltage from the output, pin 5, of the IC also changes. This is coupled

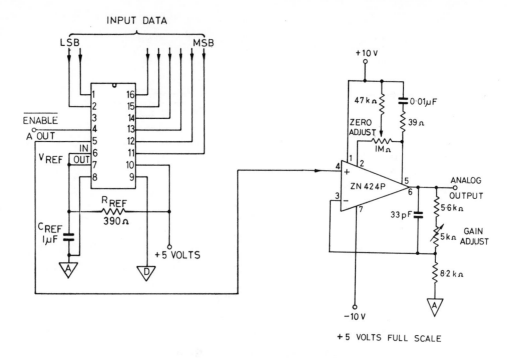

Figure 11-12. D-to-A Converter

to the input, pin 4, of a ZN424 operational amplifier. The zero adjust is set so that with all input bits low, the analog output from the op amp is 0 volts. The gain adjust in the op amp feedback loop is then set so that with all input bits high, the output is at full scale (less one LSB). Here are some sample output voltages:

Input Code (Binary)	Analog Output Voltage
11111111	4.9805V
11111110	4.9609V
11000000	3.7500V
10000001	2.5195V
10000000	2.5000V
01111111	2.4805V
01000000	1.2500V
00000001	0.0195V
00000000	0.0000V

The output is unipolar, since it varies only one direction from zero. *(Courtesy of Ferranti Semiconductors)*

ECL OR GATE

The OR gate in Figure 11-13 uses Emitter-Coupled Logic (ECL). NPN transistors Q1, Q2 and Q3 are arranged as emitter followers, and their outputs are connected together.

This circuit is a type of current mode logic, and transistors Q1, Q2 and Q3 are always turned on. When inputs A, B and C are all logical zeros (low or ground potential), the forward bias to the base of each transistor is low, emitter current is also low, and the output at D is a logical zero.

With a logical one (+ potential) to A *or* B *or* C, the transistor connected to that input is forward-biased and conducts more heavily. With increased current, the voltage drop across R4 also increases, and the output at D becomes more positive, rising to a logical one. Thus:

$$A + B + C = D.$$

Since its transistors are always on, the power consumption of ECL is even higher than that of TTL. But ECL circuitry is very fast, and can achieve propagation delay times of 1 ns or lower, and ECL flip-flops can be clocked at rates up to 400 MHz.

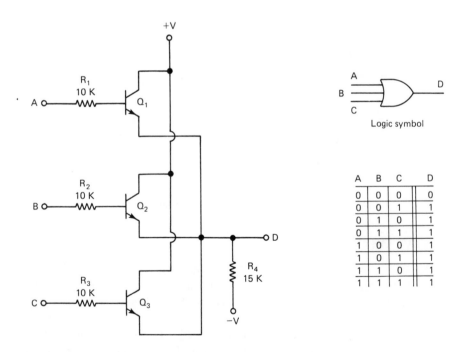

Figure 11-13. ECL OR Gate

EXCLUSIVE NOR GATE

The NOR gate in Figure 11-14 gives a logical zero output when either of the inputs is exclusively a one. However, when both inputs are either ones or zeros, the output of the gate is a logical one. This type of gate is also called an Inclusive OR gate.

Suppose the input to A is a logical one (positive voltage) and the input to B is a zero. Current flowing from B through D2 and the emitter-base junction of transistor Q1 to A causes Q1 to become forward-biased. As Q1 turns on and conducts, its collector current rises, and the voltage at D drops to a logical zero.

When the inputs to both A and B are the same (i.e., both are zeros or both are ones), there is no path for base current. Q1 is cut off, collector current falls, and the output at D rises to a logical one. The logical equation for this gate may be written as:

$$(A \cdot B) + (\overline{A} \cdot \overline{B}) = D.$$

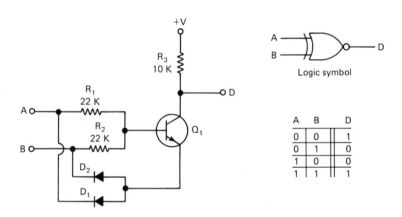

Figure 11-14. Exclusive NOR Gate

EXCLUSIVE OR GATE

The OR gate shown in Figure 11-15 is an "exclusive" OR gate—that is, the output from D is a one when either the input to A is a one or the input to B is a one, exclusively. If the inputs to A and B are the same (either ones or zeros), the output at D is a zero.

With the inputs to both A and B logical ones (+ potential), Q1 is forward-biased and turns on. Q1 collector current rises, and the voltage at the collector of Q1 becomes less positive. This, in turn, cuts off diode D4.

At the same time, however, the inputs to A and B have cut off diodes D1 and D2. With D1 and D2 cut off, diode D3 turns on, providing a path for

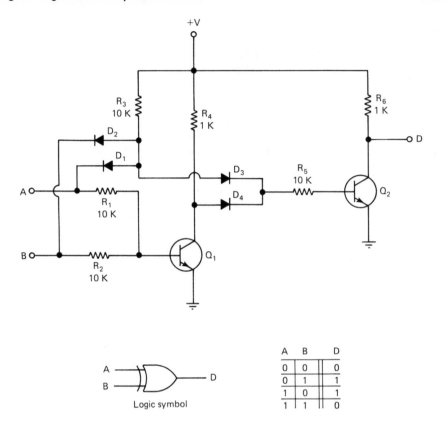

A	B	D
0	0	0
0	1	1
1	0	1
1	1	0

Logic symbol

Figure 11-15. Exclusive OR Gate

emitter-base current to forward-bias transistor Q2. As Q2 turns on, its collector current rises and the output voltage at D (Q2's collector) falls to a logical zero.

When the inputs to both A and B are logical zeros, D1 and D2 conduct and D3 is cut off. At the same time, however, Q1 is cut off, so its collector current falls, and the voltage at the anode of D4 arises to a point where D4 turns on. This, too, provides a path for Q2 base current, forward-biasing Q2 and giving a low voltage at the collector of Q2. So the output at D is again a logical zero.

When the input at A is a logical one and the input at B is a logical zero, Q1 is forward-biased by input A and conducts, causing D4 to be cut off. At the same time, the low input at B causes D2 to conduct, cutting off diode D3. With both D3 and D4 cut off, there is no path for base current for Q2, and that transistor, too, is cut off. Collector current through Q2 falls, and the output voltage at D rises to a logical one. The logical equation for the Exclusive Or gate is:

$$(A \cdot \overline{B}) + (\overline{A} \cdot B) = D.$$

FLIP-FLOP

A flip-flop is a memory circuit—it can actually "remember" the last input it received. In the flip-flop shown in Figure 11-16(a), if the SET switch is closed momentarily, the base current through Q2 turns that transistor on, resulting in large collector current and $\overline{X}$ falling to a logical zero. At the same time, that logical zero is felt at the base of Q1, ensuring that *that* transistor remains cut off and that the output at X is a logical one. Since X is a logical one (positive voltage), Q2's base current is maintained through resistor R6 and Q2 remains turned on.

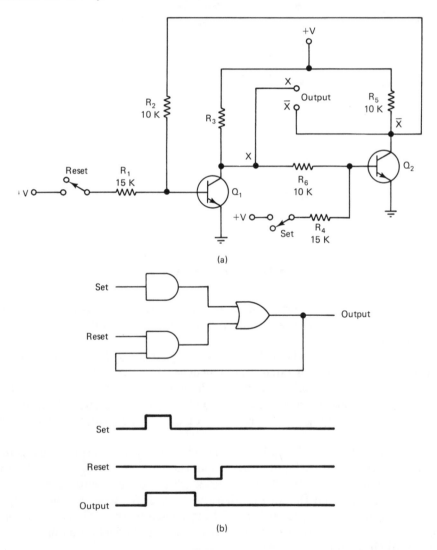

(a)

(b)

Figure 11-16. Flip-Flop

If the SET switch is closed repeatedly—simulating a series of digital inputs—nothing changes in the flip-flop. However, if the RESET switch is thrown, transistor Q1 turns on, the output at X falls to a logical zero, and Q2 is cut off. With Q2 cut off, the potential at X̄ rises to a logical one, and remains there until the flip-flop is "toggled" again.

Figure 11-16(b) shows a flip-flop made up of two AND gates and an OR gate.

HIGH-SPEED TTL NAND GATE

The TTL NAND gate shown in Figure 11-17 is similar to an ordinary TTL NAND except that the resistor values are somewhat lower and transistors Q3 and Q4 have been added in a Darlington configuration. The added circuitry increases the current switching speed of Q4, and this, together with the lower resistance values, decreases the length of time it takes this gate to change states. High-speed TTL such as this has a propagation delay of 6 ns, consumes 22 mW of power per gate, and has an operating frequency for flip-flops of up to 50 MHz.

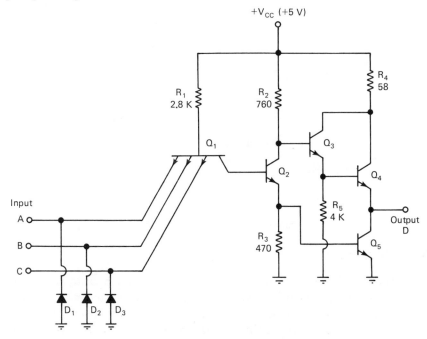

Figure 11-17. High-Speed TTL NAND Gate

HNIL NAND GATE

High Noise Immunity Logic, or HNIL (also called High Threshold Logic, or HTL) is necessary in many industrial and control applications for digital circuitry, where noise levels and voltage transients are often far above those encountered in computer and similar environments. HNIL is also used in telephone switching, SCR, and line driving circuits.

Figure 11-18 shows a HNIL NAND gate. The key to the gate's ability to ignore noise is D3, a 5.8V Zener diode which raises the input threshold and gives the gate high noise immunity.

When the inputs to both A and B are logical ones (+ potential), D1 and D2 are cut off and D3 goes into Zener operation, permitting base current in Q1. This forward-biases and turns on Q1, resulting in a logical zero at output C.

The logic levels necessary to cause HNIL to switch are normally much higher than those of other logic families. Noise immunity of the gate shown is 3.5V with 12V V_{CC}. (Compare this with typical TTL noise immunity of 0.4V.) HNIL is the slowest of any logic family—propagation delay time is about 150 ns—and consumes the most power.

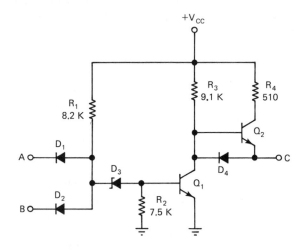

Figure 11-18. HNIL NAND Gate

INVERTER CIRCUIT

Figure 11-19 illustrates an inverter circuit. With no input to A, transistor Q1 is cut off and the positive potential at its collector causes a logical one to be the output through resistor R3. When a logical one (positive potential) is the input at A, the current through the base of Q1 causes the transistor to conduct, resulting in a large collector current and the dropping of the output at B to a logical zero. The inverter is also called a NOT circuit, since A = $\overline{B}$ (that is, A yields NOT B).

LED DRIVER

The displays for electronic watches, calculators, and electronic test equipment often employ light-emitting diode segments. Figure 11-20 illus-

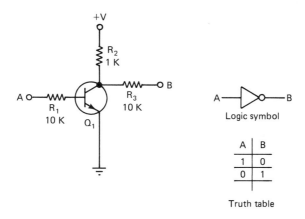

Logic symbol

A	B
1	0
0	1

Truth table

Figure 11-19. Inverter Circuit

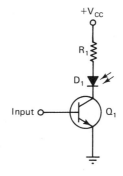

Figure 11-20. LED Driver

trates an LED driver. When the input to the base of Q1 is a logical one (becomes more positive), Q1 is forward-biased and collector current flows, turning on LED D1. Resistor R1 is selected for the desired operating current through the LED.

LLL NOR GATE

The LLL (Low-Level Logic) NOR gate shown in Figure 11-21 responds to input levels that are only very slightly differentiated between zeros and ones. This is possible because a positive input, a one, at A, B, or C is not an active part of Q1's base circuit, but only causes the base diode D4 to switch.

With the inputs to A, B and C set at logical zeros (that is, the inputs are low) D1, D2 and D3 are cut off and D4 conducts. The potential at the base of Q1 is not positive enough to forward-bias the transistor, so Q1 is cut off and the collector voltage, and the output at D, are high.

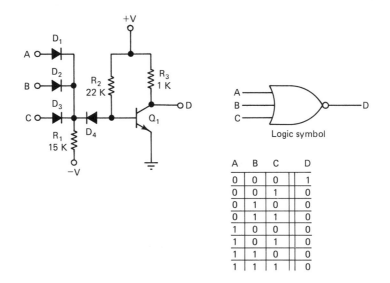

A	B	C	D
0	0	0	1
0	0	1	0
0	1	0	0
0	1	1	0
1	0	0	0
1	0	1	0
1	1	0	0
1	1	1	0

Figure 11-21. LLL NOR Gate

Even a slight positive potential, a logical one, input to A, B, or C is enough to turn on the diode (D1, D2, or D3) associated with that input. In turn, the potential at the cathode of D4 becomes more positive, cutting off that diode and raising the potential at its anode to a level that is now sufficient to forward-bias Q1. Q1 turns on, its collector current rises, and the voltage at D falls to a logical zero. Thus, for this as for other NOR gates:

$$A + B + C = \overline{D}.$$

NMOS NAND GATE

NMOS circuitry consists of N-channel MOSFETs that have been diffused onto an integrated circuit chip. In the NMOS NAND gate shown in Figure 11-22, when the inputs to A and B are both low, only the load device, Q1, is on. Both Q2 and Q3 are off, and the output at C is a logical one (+ potential). Since the FETs are in series, as long as the input to the gate of either Q2 or Q3 is a logical zero there will be no current flow, and the output at C will remain a one.

When both A and B are logical ones, both Q2 and Q3 are forward-baised by the positive potentials at their gates. With Q2 and Q3 on, drain current flows through all three transistors, causing the output at C to drop to a logical zero. Thus:

$$A \cdot B = \overline{C}.$$

NMOS transistors have about one-third the resistance of PMOS devices, so an NMOS circuit, although more difficult and expensive to manufacture, is smaller in size, switches faster, and uses less power than an equivalent PMOS circuit.

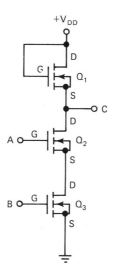

Figure 11-22. NMOS NAND Gate

NOR CIRCUIT

In the simple NOT OR, or NOR, circuit shown in Figure 11-23, with a logical zero applied to both input A and input B, transistor Q1 is cut off and

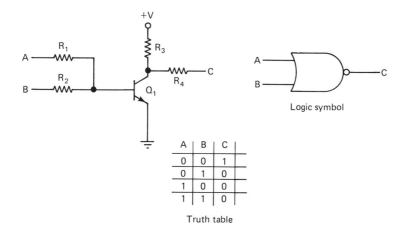

A	B	C
0	0	1
0	1	0
1	0	0
1	1	0

Truth table

Figure 11-23. NOR Circuit

the output at C is a logical one. However, if a logical one (positive potential) is applied to A *or* B, Q1 conducts and the output at C falls to a logical zero. The logic equation for this circuit is:

$$\overline{A + B} = C.$$

The logic diagram and logic symbol for the circuit are also shown.

PMOS NAND GATE

In hand-held calculators, aerospace equipment, and other electronic applications requiring digital circuits with small size and low power consumption, PMOS circuitry, because of its ease of manufacture and reliability, was the first technology to be widely used.

Figure 11-24 shows a NAND gate consisting of three P-channel MOS-FETs. When the input to the gate of either Q2 or Q3 is a logical zero, that field-effect transistor is cut off, little drain current can flow, and the output at the drain of Q2 is a logical one (that is, a − potential). Q1 is always on, and acts like a load resistor in this circuit.

When the inputs at A *and* B both become ones (become more negative), the − voltages at the gates of Q2 and Q3 turn these transistors on. Drain current then flows freely through the source-drain regions of Q2, Q3, and load transistor Q1. The output at C becomes less negative, falling to a logical zero. Thus:

$$A \cdot B = \overline{C}.$$

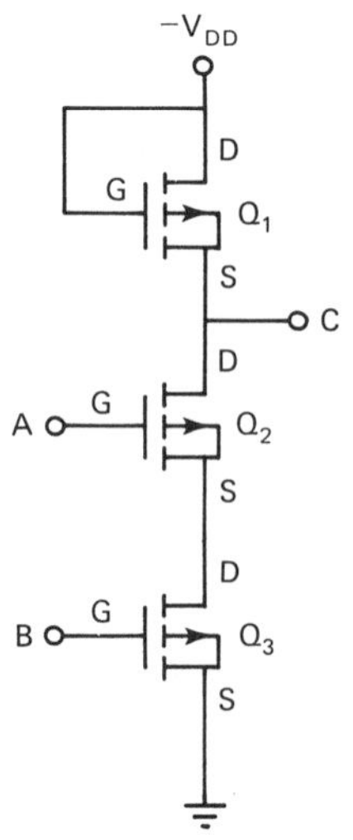

Figure 11-24. PMOS NAND Gate

RTL NAND GATE

In RTL, or Resistor-Transistor Logic, transistors are used as the main switching elements. The NAND gate of Figure 11-25 uses three NPN transistors connected with their emitters in series. When a logical zero is input to the base of any of these transistors from A, B or C, that transistor is cut off and the voltage at the collector of Q1 remains at a logical one (high). However, when A *and* B *and* C are all logical ones (that is, + voltages), Q1, Q2 and Q3 are forward-biased, and current flows through their collectors. This causes the potential at the collector of Q1 (and, hence, at output D) to fall to a logical zero. Thus:

$$A \cdot B \cdot C = \overline{D}.$$

RTL was the first logic family to be established as a commercial product line. RTL can achieve gate propagation times of 12 ns and can consume as little as 10 mW of power per gate, but RTL gates are susceptible to noise and their fanout capability is low.

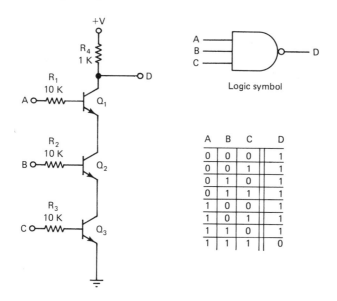

A	B	C	D
0	0	0	1
0	0	1	1
0	1	0	1
0	1	1	1
1	0	0	1
1	0	1	1
1	1	0	1
1	1	1	0

Figure 11-25. RTL NAND Gate

RTL NOR GATE

The RTL NOR gate in Figure 11-26 uses Resistor-Transistor Logic to accomplish its logic function. With a logical zero (no or low voltage) at each of the three inputs, Q1, Q2 and Q3 are cut off, resulting in a logical one at output D. If the positive voltage of a logical one appears at A *or* B *or* C, however, the emitter-base junction of the NPN transistor (Q1, Q2 or Q3) connected to that input is forward-biased. Current through this junction in turn forward-biases the emitter-collector junction, causing high collector current; the voltage at D falls to a logical zero. Thus:

$$A + B + C = \overline{D}.$$

Resistor-Transistor Logic is among the oldest and most flexible forms of digital logic available.

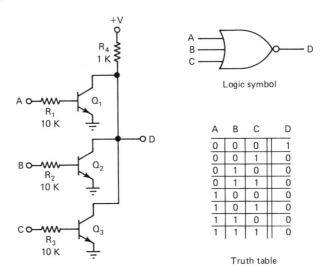

A	B	C	D
0	0	0	1
0	0	1	0
0	1	0	0
0	1	1	0
1	0	0	0
1	0	1	0
1	1	0	0
1	1	1	0

Truth table

Figure 11-26. RTL NOR Gate

SCHOTTKY TTL NAND GATE

Figure 11-27(a) shows an example of a Schottky TTL (Transistor-Transistor Logic) NAND gate. The gate operates like a normal TTL NAND circuit except that it uses Schottky clamping on most of its transistors to decrease switching time.

In the circuit, Q1, Q2, Q3, Q4 and Q5 are all symbols for Schottky-clamped transistors. Such a transistor is formed by depositing a fine layer of metal that overlaps the N-type collector region and the P-type base region of an NPN transistor. This layer of metal helps form a *Schottky barrier diode* (SBD) clamp between the base and the collector, as shown in Figure 11-27(b).

In operation, when a positive input to the base turns on a Schottky-clamped transistor, base current drives the device towards saturation. As

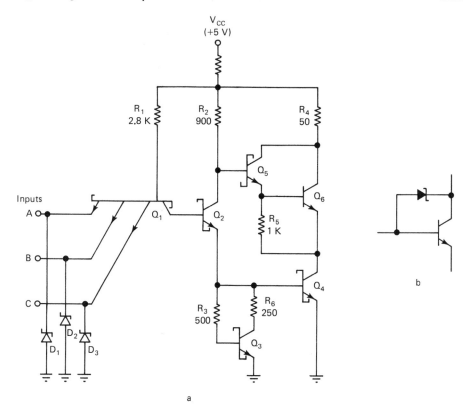

Figure 11-27. Schottky TTL NAND Gate

the collector current rises, however, and collector voltage drops, the SBD begins to conduct, removing excess base current and keeping the transistor out of deep saturation. With the excess base charge not stored, switching time is reduced markedly, and Schottky TTL is much faster than ordinary TTL. Circuitry of this type consumes 19 mW of power per gate, has a propagation delay of only 3 ns, and can switch at a maximum flip-flop frequency of 125 MHz.

TTL NAND GATE

TTL (Transistor-Transistor Logic) or T^2L is one of the most widely used forms of digital logic. In the NAND gate shown in Figure 11-28, inputs to the gate are fed through Q1, an NPN transistor with multiple emitters. Q2 is a switching transistor or phase splitter, while Q3 and Q4 are a totem-pole or active pull-up output stage.

In operation, Q1 is always turned on. When any of the three inputs is a logical zero, current flowing through R1 is diverted to that emitter line,

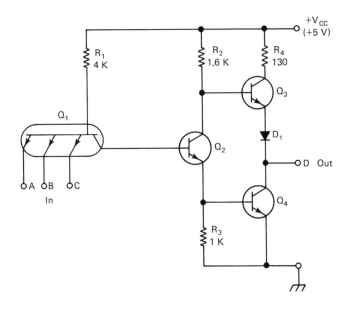

Figure 11-28. TTL NAND Gate

transistor Q1 is saturated, and there is insufficient base current to Q2 to forward-bias that transistor. With Q2 cut off, the voltage at the collector of Q2 is near $+ V_{CC}$, so Q3 is forward-biased and turned on. Q4 is off, the voltage at the collector of Q4 is high, and the output felt at D is a logical one ($+$ potential near V_{CC}).

When the inputs to A *and* B *and* C are all logical ones, Q1 remains on but the current through R1 is now shunted through Q1's collector region into the base of transistor Q2. With this increased base current, Q2 is quickly turned on and provides base current for Q4, turning on that transistor also. The voltage at the collector of Q2 drops and Q3, no longer forward-biased, turns off. Since Q4 is on and fully saturated and Q3 is off, the output at D falls to a logical zero. Thus

$$A \cdot B \cdot C = \overline{D}.$$

Normal TTL has a typical gate delay of 10 ns, power dissipation of 10 mW per gate, and a maximum flip-flop operating frequency of 35 MHz.

CHAPTER 12

Power Supplies

INTRODUCTION

Power supplies are truly the workhorses of the electronics world. Often unsung, indicated on a schematic diagram only by a reference to "V_{cc}," "-24V PTT," or other cryptic allusion, power supply circuits range from the very simple to the extremely complex. Because a single circuit must handle all the power consumed by hundreds or thousands of other components, power supply circuitry often takes up the largest physical area in an instrument, calculator, or other device. Power supplies are usually mounted so that their often considerable heat can be dissipated.

The first duty of a power supply often is power *conversion*. This may include *rectification*, where alternating current is changed to DC for use in supply voltages. It can also be DC-to-DC conversion, where a DC voltage is reduced or stepped up to a desirable value. Or the power supply may even perform DC-to-AC conversion, changing the output from a battery to a source of alternating current.

After alternating current is rectified, it becomes pulsating direct current; although it is DC, strictly speaking, it nevertheless varies in amplitude. This variation, or ripple, can occasionally be tolerated, but usually it is necessary to remove it and to create relatively pure direct current. So another task of a power supply is *filtering*, where the pulsating DC is smoothed out and much of its ripple component removed. Filtering is usually performed by RC, RL, or LC combinations in various types of circuits connected at the output of the rectifier.

Another function of power-supply circuitry is *regulation*. Load, input voltage, temperature, or other factors can act to change the output from a power supply. However, regulator circuits are often included in power supplies to provide relatively constant voltage and/or current over a wide range of conditions.

A regulator circuit is essentially a feedback loop. As the output voltage or current changes, the change is fed back to a pass element (usually a transistor or SCR) that passes more or less of the input as needed to maintain a constant output. The pass element may be operated linearly or at cutoff or saturation; when the pass element is operated so it goes into cutoff or saturation, the circuit is called a switching regulator.

Linear regulators may be either series or shunt configurations, depending on whether the pass element is connected in series or in parallel with the output.

Hand-in-glove with the regulatory function of a power supply is *protection*. Power supplies are often designed for no-load to full-load regulation, and many also have additional circuitry to prevent damage to the power supply itself in the event of an external short circuit.

AUTOMATIC NICAD BATTERY CHARGER

The circuit shown in Figure 12-1 is of an experimental fast charger for nickel-cadmium batteries. Nicad batteries can accept pulsed charging current

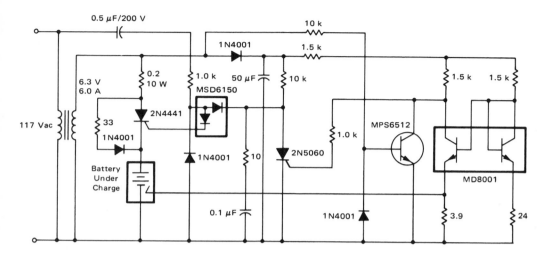

Figure 12-1. Automatic Nicad Battery Charger

better than steady direct current, and this characteristic enables the use of a pulsed 60 Hz charger that is driven from ordinary line current.

Charging current to the battery is supplied from the secondary of the transformer through the 0.2-ohm, 10 watt current-limiting resistor and the 2N4441 SCR. The trigger signal for the SCR is produced by the charge and discharge of the 0.5 μF/200V capacitor connected to the transformer primary. When the SCR is cut off, the 33-ohm resistor and 1N4001 diode provide a trickle charge for the battery.

This charger uses "third electrode" sensing to detect when the fast-charge cycle is complete. The third electrode is actually a part of the battery; although the voltage from this electrode is small, its potential at full charge is enough to cause the left-hand transistor in MD8001 to cut off, raising its collector voltage. This sends a shutoff signal to the 2N5060 SCR, and this, in turn, causes the 2N4441 SCR to turn off. The diodes in the MSD6150 network prevent interaction of the charge current and the shutoff signal. (*Courtesy of Motorola, Inc.*)

CASCADE VOLTAGE MULTIPLIER

Many stages of the half-wave voltage doubler can be stacked together, as shown in Figure 12-2. Depending on the number of stages, this type of circuit can provide voltage doubling, quadrupling, etc. The particular circuit shown multiplies the average voltage across the transformer secondary eight times. Since each stage is a voltage doubler, the final capacitor should have a voltage rating that is twice the final output voltage. Typically, all capacitors have a capacitance of 20-50 μF.

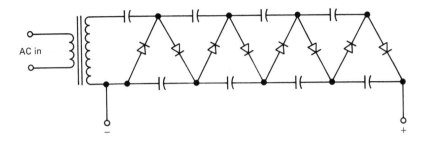

Figure 12-2. Cascade Voltage Multiplier

CONSTANT CURRENT SOURCE

Some electronics circuits require the current through them to be constant even though the amount of load varies. Figure 12-3 illustrates how an NPN transistor is used as a source of constant current.

Q1 is connected as an emitter follower. The voltage at the emitter, then, is the same as the voltage at the base. And the voltage at the base is set at about one volt by the divider consisting of R1 and R2.

Since the voltage at the emitter is constant, the emitter current (and hence, collector current) is determined by emitter resistor R3. The collector current will remain constant over a wide range of loads.

This type of circuit is often found in difference amplifiers and operational amplifiers, where a source of constant current is necessary to achieve a good common-mode rejection ratio.

The circuitry for which this configuration provides constant current is connected in place of load resistor R_L.

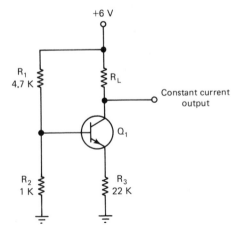

Figure 12-3. Constant Current Source

CROWBAR CIRCUIT

Figure 12-4 illustrates how a silicon-controlled rectifier is utilized as a "crowbar" in order to provide overvoltage protection and regulation in a power supply.

PNP transistor Q1 functions as a voltage comparator and amplifier. Zener diode Z1 provides a reference voltage at the emitter of Q1, and R3 is set so the transistor is normally turned off. If a transient or other high-voltage spike causes the voltage at V+ to rise, base current flows in Q1 and the transistor turns on. Collector current drawn through Q1 and R5 triggers the gate of SCR1, and the SCR fires, appearing as a fast-acting short circuit to the power supply at V+. R6 and C1 form a low-impedance RC network to filter out noise spikes that might appear on the gate of SCR1.

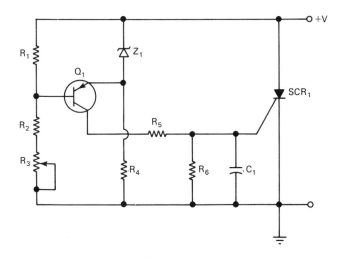

Figure 12-4. Crowbar Circuit

FULL-WAVE BRIDGE RECTIFIER

By using four diodes in a bridge circuit, as shown in Figure 12-5, full wave rectification is achieved without the necessity for a center-tapped transformer. When the voltage at point A of the transformer secondary swings positive, the path for current flow is through diode CR3, through load resistor R_L, and through diode CR2. On the other half-cycle, when point A swings negative, current flows through CR1, R_L, and CR4. The pulsating DC output across R_L, then, is the same as for any full-wave rectifier.

In practice, many manufacturers prepackage such a diode bridge, encapsulating all four rectifier diodes in a single case.

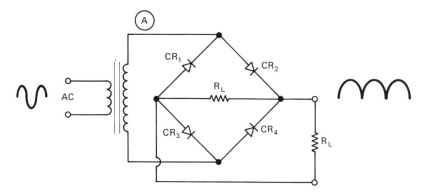

Figure 12-5. Full-Wave Bridge Rectifier

FULL-WAVE RECTIFIER

A basic full-wave rectifier that incorporates two diodes and a center-tapped transformer is shown in Figure 12-6. When point A at the transformer secondary swings positive with respect to the center tap, diode CR1 is forward-biased and conducts, causing an IR drop across load resistor R_L to develop the output voltage. Since point B is negative with respect to the center tap at this time, CR2 is reverse-biased and cut off.

On the opposite half-cycle of AC input, point B becomes positive with respect to the center tap, while point A swings negative. At this time, CR2 conducts, CR1 is cut off, and the output voltage is again developed across R_L.

This type of circuit rectifies both halves of each input cycle; the output is pulsating direct current. Assuming no voltage drop across CR1 or CR2, a voltmeter placed across R_L will read the *average* voltage present, which is 0.9 times the RMS voltage between the center tap and one end of the transformer secondary (or 0.636 times the peak voltage at the same point).

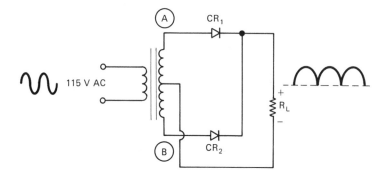

Figure 12-6. Full-Wave Rectifier

FULL-WAVE VOLTAGE DOUBLER

The circuit in Figure 12-7 uses capacitors to double the rectified output voltage from the secondary winding of T1. During the half-cycle when diode D2 is conducting, capacitor C2 charges to the peak voltage. During the opposite half-cycle, C1 is charged up by current through D1 and by the discharging of C2. Thus, the average voltage across R_L is twice the amplitude of the average potential at the secondary of T1.

C1 and C2 must both be rated higher than the maximum voltage across the secondary of T1. Because its regulation is poor, this type of circuit is usually found only in low-power applications.

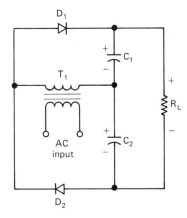

Figure 12-7. Full-Wave Voltage Doubler

HALF-WAVE RECTIFIER

Figure 12-8 shows how a semiconductor diode is used to form a basic half-wave rectifier circuit. When the voltage induced at point A of the transformer secondary is positive, diode CR1 is forward-biased and readily permits current flow. The resulting IR drop across load resistor R_L provides an output voltage at this time.

During the other half-cycle, when point A swings negative, CR1 is reverse-biased and cut off, so no current flows. The output of the half-wave rectifier, then, is pulsating DC; that is, direct current that varies from 0 volts to some positive level.

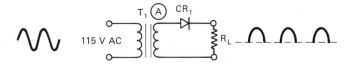

Figure 12-8. Half-Wave Rectifier

A DC voltmeter placed across R_L will read the *average* value for voltage at the output (assuming no voltage drop across CR1). This is 0.45 times the AC root-mean-square voltage, or 0.318 times the peak voltage, present across the transformer secondary.

HALF-WAVE VOLTAGE DOUBLER

A capacitor in series with the transformer secondary winding is one identifiable characteristic of the half-wave voltage doubling circuit, or cascade doubler, illustrated in Figure 12-9. During one half-cycle of input, series capacitor C1 charges through diode D1. On the other half-cycle, C2 charges through D2 and C1. Within a few cycles, C2 is charged to an average value that is about twice the average voltage available at the secondary of T1.

The cascade doubler has one side of its output in common with one side of the transformer secondary. However, for the same size capacitors, it has higher ripple and poorer regulation than does the full-wave doubler.

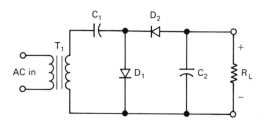

Figure 12-9. Half-Wave Voltage Doubler

HIGH-CURRENT VOLTAGE REGULATOR

The adjustable voltage regulator shown in Figure 12-10 uses three RCA 2N2016 power transistors in parallel as the series pass regulating element, giving the circuit an output of up to 10A at 30 VDC. This circuit provides line regulation within 1.0% and load regulation with 0.5%. (*Courtesy of RCA Solid State.*)

JENSEN DC-TO-DC CONVERTER

The Jensen circuit shown in Figure 12-11 uses two transistors and two high-frequency transformers to form a square-wave oscillator that, in a manner similar to the operation of the Royer oscillator, acts as a DC-to-DC converter. The windings of the left-hand transformer are opposite in polarity; this transformer also has a square hysteresis loop characteristic, which permits it to change from a saturated condition in one direction to a saturated condition in the opposite direction very quickly. It is the hysteresis characteristic of the transformer that primarily determines the operating frequency of the oscillator.

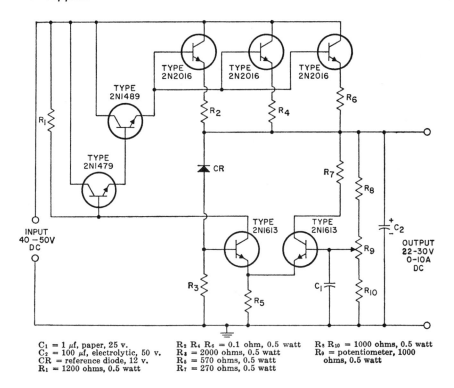

C₁ = 1 μf, paper, 25 v. R₂ R₄ R₆ = 0.1 ohm, 0.5 watt R₈ R₁₀ = 1000 ohms, 0.5 watt
C₂ = 100 μf, electrolytic, 50 v. R₃ = 2000 ohms, 0.5 watt R₉ = potentiometer, 1000
CR = reference diode, 12 v. R₅ = 570 ohms, 0.5 watt ohms, 0.5 watt
R₁ = 1200 ohms, 0.5 watt R₇ = 270 ohms, 0.5 watt

Figure 12-10. High-Current Voltage Regulator

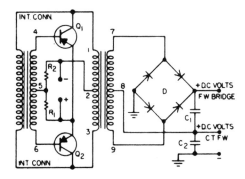

Figure 12-11. Jensen DC-to-DC Converter

In the Jensen circuit, the right-hand (output) transformer does not go into saturation, and this prevents a collector current spike. At the output of the right-hand transformer is a full-wave bridge rectifier.

The DC-to-DC converter is often found in portable and mobile equipment, where a low battery voltage must be boosted to some more useful level. (*Courtesy of Microtran Company, Inc.*)

RINGING CHOKE DC-TO-DC CONVERTER

In the basic ringing choke DC-to-DC converter shown in Figure 12-12, a blocking oscillator (or chopper) is transformer-coupled to a half-wave rectifier that converts the pulsating oscillator output into direct current.

When power is applied to PNP transistor Q, its collector current rises, conducting through the top primary winding of the transformer, until the transistor goes into saturation. Base drive from the lower primary winding aids this process. Diode D in the secondary of the transformer is wired so that no power is delivered to the load at this time.

With the transistor in saturation, collector current through the primary of the transformer no longer continues to rise, and the lower primary winding can no longer maintain enough base drive to keep Q saturated. As the current through it decreases, the voltage induced in this feedback winding rapidly decreases, driving Q beyond cutoff. At this time, the energy stored in the primary of the transformer is released, and the collapsing magnetic field induces a voltage in the secondary. This voltage is now coupled through diode D to filter capacitor C2 and the load. C2 helps maintain a constant output voltage even when there is no output from the oscillator. (That is, when Q is conducting.)

This type of circuit is not very efficient and has a high ripple component in its output; however it is found in many low-power applications because it is simple and uses few components. (*Courtesy of Microtran Company, Inc.*)

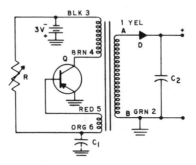

Figure 12-12. Ringing Choke DC-to-DC Converter

ROYER OSCILLATOR DC-TO-DC CONVERTER

A basic single-transformer DC-to-DC self-oscillating converter is shown in the Royer oscillator of Figure 12-13(a). By using for T1 a transformer with the hysteresis curve shown in Figure 12-13(b), the drive to the bases of Q1 and Q2 is a relatively good square wave, which results in an efficient converter.

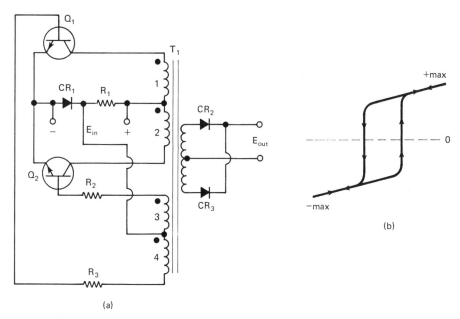

(a)

(b)

Figure 12-13. Royer Oscillator DC-to-DC Converter

In the schematic diagram, the dots on the transformer windings indicate that the windings are parallel; that is, if the polarity at the dot of a winding is positive relative to the undotted end, it is positive at the dots on the other windings as well.

To analyze this circuit, consider Q1 to be turned on and in saturation, and the potential at the dot ends of the windings to be negative. Thus, collector current flows through winding 1 of T1. The potential across winding 4 biases the base of Q1 negative with respect to its emitter, holding Q1 on. At the same time, Q2 is held off by windings 2 and 3.

Q1 remains on as long as there is a voltage induced in winding 1 and the hysterisis loop of T1 is climbing. At +MAX, the transformer core saturates, and the voltages across the primary windings fall to zero. This forces the collector of Q1 towards $+E_{IN}$, turning the transistor off.

As the lines of flux in the transformer collapse, they induce voltage in the opposite direction: the dot ends of the windings are now positive, and Q2 is turned on and quickly saturates. The hysteresis curve of T1 builds again to −MAX.

This process continues, generating in the secondary of the transformer a square wave that is rectified by CR2 and CR3. In a step-up DC-to-DC converter, a voltage doubler or multiplier is often used in the secondary of T1 to boost the voltage even more.

The Royer oscillator has a large spike of collector current each time the transformer goes into saturation. The Jensen circuit uses a second, non-saturating, transformer to avoid this spike.

SERIES REGULATOR

In a series regulator, the pass element is actually in series with the output. Figure 12-14 shows a simple series regulator. The voltage at the base of the emitter follower (NPN transistor Q1) is held constant by the reference element, Zener diode Z1. The conduction of Q1 changes to compensate for any swing, positive or negative, of V_{OUT}.

The series regulator responds faster and is more efficient than the shunt regulator. However, since the pass element is in series with the output, the transistors used in series regulators must usually have very healthy power ratings.

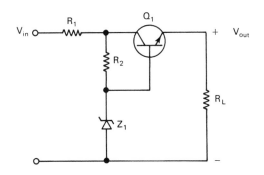

Figure 12-14. Series Regulator

SERIES REGULATOR WITH CURRENT LIMITING

One of the problems of a series regulator is that if the output is short-circuited, all the available supply current and voltage is applied through the series-pass element. If the transistor is not rated to dissipate all the power, it will quickly burn out. Because the thermal time constant of a fuse is much longer than that of many transistors, a fuse placed in the line does not always protect the pass element in a series regulator.

A solution used in many power supply circuits is current limiting. In Figure 12-15, current limiting is provided by diode CR1 and transistor Q4. Q1 is the series-pass element, connected in a Darlington configuration with Q2.

In operation, the base-emitter voltage of Q1 and Q2 is proportional to the output current of the circuit. During an overload condition, these voltages rise, causing Q4 and CR1 to conduct. Q4 shunts a portion of the bias that is normally available for the regulator transistor, and the series resistance of Q1 is effectively increased.

Temperature drift in the circuit is minimized by mounting Q1 and Q4 on a common heat sink.

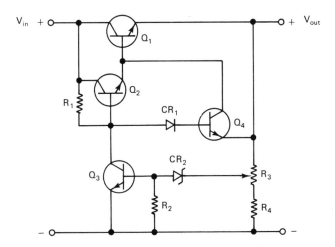

Figure 12-15. Series Regulator with Current Limiting

SHUNT VOLTAGE REGULATOR

The basic shunt regulator is made up of a reference-voltage element and a shunt element. In Figure 12-16, the reference voltage is selected via Zener diode Z1, while the shunt element is NPN transistor Q1, connected in a common-emitter configuration.

Changes in the input voltage are regulated by shunt transistor Q1. For example, if the output voltage rises, more collector current flows through Q1, absorbing the rise and maintaining the output voltage at the level selected by Z1 and R1.

Shunt regulators are much simpler than series regulators, but are not as efficient.

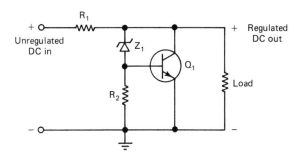

Figure 12-16. Shunt Voltage Regulator

STEP-UP DC-TO-DC REGULATOR

An LM380 audio amplifier integrated circuit chip is used as an oscillator in the DC-to-DC regulator shown in Figure 12-17. The oscillator changes +14VDC into an 8-volt peak-to-peak square-wave signal.

The amplitude of this square wave is stepped up to 50 volts across transformer T1 and the voltage doubler consisting of CR1, CR2, C8 and C9. Series-pass regulator Q1 provides a regulated 38 VDC to the input of the 78L24 IC chip, which reduces the output to 24 VDC and provides further regulation. (*Courtesy of Wulfsberg Electronics, Inc.*)

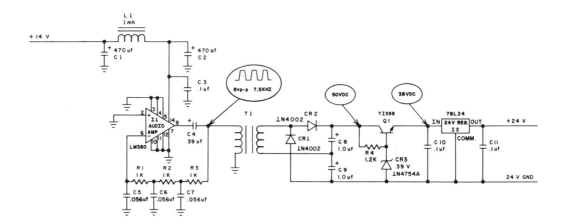

Figure 12-17. Step-Up DC-to-DC Regulator

SWITCHING POWER SUPPLY PRINCIPLES

Figure 12-18 illustrates the principles of a switching power supply. This type of configuration is used for both power conversion and regulation.

A series-pass element, emitter-follower transistor Q1, is placed in series with the input voltage. The transistor is not operated linearly, but instead is switched on and off by a square wave applied to its base. By varying the length of time the square wave holds the transistor on, the voltage through it is controlled.

By using square waves to control series or shunt elements, power supplies can be operated at relatively high frequencies (up to 100 kHz), which makes filter and other components very small physically. Switching regulators and converters are used in all types of step-up and step-down AC-to-DC, DC-to-DC, and DC-to-AC circuitry because of their small size and weight, good regulation, and low internal losses.

Figure 12-18. Switching Power Supply Principles

CHAPTER 13

Radio and Television Receivers

INTRODUCTION

A radio or television receiver is an example of the use of many different circuits in a complete electronic device. The receiver must pick up the desired transmitted signal, while rejecting all other signals. It must amplify this signal to a useful level, then it must use a detector to separate the modulating component from the RF signal. Finally, it must amplify the resulting audio or picture signal still further before applying it to the speaker or picture tube. Naturally, a power supply, often capable of producing several different voltages simultaneously, is also needed.

AM Radio Receivers

The simplest AM receiver is the crystal detector, in which there is no amplification of the signal before or after detection. But adding positive feedback from the output of a detector to its input dramatically increases the sensitivity and improves its selectivity as well. This is known as the regenerative detector; good selectivity is achieved by injecting negative feedback into the tuned circuit, thus improving its Q (figure of merit).

Maximum gain can be obtained from a single tube or transistor by using the super-regenerative circuit. This type of detector is almost as sensitive as many multistage receivers. It is an adaptation of the regenerative detector; the detector swings in and out of oscillation at a rapid rate using a self-squegging or an external quench oscillator. The self-squegging super-regenerative circuit is widely used in pocket radio paging receivers where small size and high selectivity are required.

The TRF receiver is similar to a simple detector circuit, but several stages of RF amplification are added before detection. This type of receiver was very popular in the 1930s, when four-tube table model radios were sold for less than $15. These receivers had adequate sensitivity for local reception, but as more and more radio stations began broadcasting, there came a need for better selectivity. This was achieved by down-converting the frequency of an intercepted radio signal to a lower frequency at which more amplification and greater selectivity could be obtained.

Most modern AM radio receivers use some version of the superheterodyne circuit. The superhet receiver was invented by Major Edwin H. Armstrong, who also invented FM and the super-regenerative receiver. The superhet provides very high gain and excellent sensitivity.

As shown in Figure 13-1, the received RF signal is coupled first to a converter, which changes the signal to a lower intermediate frequency (IF). The IF passes through one or more stages of amplification, then the modulation is removed from the signal by a detector. After detection, the audio signal is amplified still further before it is applied to the speaker.

FM Receivers

Most FM receivers also use a version of the superheterodyne circuit. However, as shown in Figure 13-2, an additional limiting stage may be added immediately before the detector. This IF limiter removes

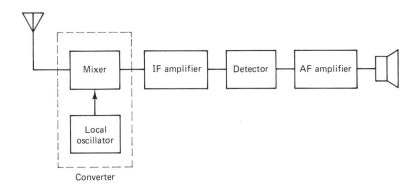

Figure 13-1. AM Superheterodyne Receiver

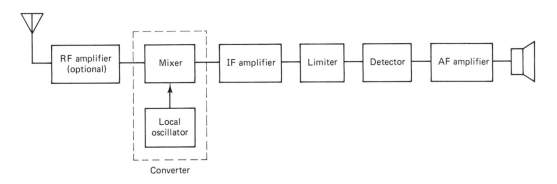

Figure 13-2. FM Superheterodyne Receiver

any amplitude modulation and feeds the detector a constant amplitude signal whose frequency changes.

In the typical FM receiver, the RF signal is boosted by an amplifier, then fed to a mixer stage, where it is converted to an intermediate frequency. Let's assume that the receiver is tuned to 53 MHz and the IF is 10.7 MHz. The local oscillator (LO) may operate at 63.7 MHz or 42.3 MHz, since both of these frequencies are 10.7 MHz removed from 53 MHz. When the 53 MHz signal is fed to the mixer, it combines with the LO signal, and a beat frequency at 10.7 MHz is generated. This 10.7 MHz IF signal is fed to the IF amplifier. After amplification, it is fed to the detector, and the resulting audio is passed on the audio amplifier and speaker. Limiter stages eliminate amplitude variations in the signal and allow only frequency variations to be passed along to the detector.

The double conversion superheterodyne utilizes two mixers and two local oscillators. The added stages are shown in Figure 13-3. In

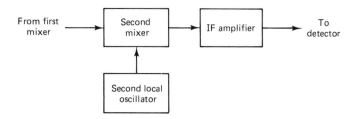

Figure 13-3. Added Stages for Double Conversion Receiver

a typical double-conversion VHF receiver, the signal from the 150 MHz VHF band is down-converted to 10.7 MHz by the first mixer and then is down-converted still further to 455 kHz by the second mixer.

Because FM communications receivers are designed to receive narrow-band FM signals (± 5 kHz), their selectivity must be considerably better than that of an FM broadcast band receiver, which is designed to pass FM signals with a deviation of ± 75 kHz.

Television Receivers

A television signal is a particularly complex one, carrying as it does information for both TV sound and a black-and-white (and in most cases, color) picture. Shown in Figure 13-4 is a diagram of the

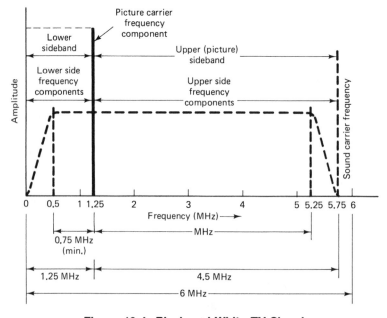

Figure 13-4. Black-and-White TV Signal

type of signals used in black-and-white transmission; it is known as a vestigial-sideband transmission because most (but not all) of the lower sideband is suppressed.

The composite signal bandwidth is 6 MHz. The picture carrier frequency is indicated by the tallest line; its upper sideband is 4 MHz wide and contains picture information. The lower sideband also contains picture information, but is only 1.25 MHz wide. Thus, only a part of the lower sideband remains, giving rise to the name vestigial-sideband AM transmission.

The sound carrier for the channel is located 4.5 MHz above the center frequency of the picture carrier. This sound carrier is an FM signal with a bandwidth of 0.5 MHz.

Color TV transmission is still more complex. All colors can be broken down into various quantities of red, green and blue. And these three colors can, in turn, be represented by phase relationships between two signals, the so-called Q and I signals. For a color TV signal, a 3.58 MHz chrominance subcarrier is added to the picture sideband before transmission. Q and I signals containing the color information are placed onto this 3.58 MHz subcarrier, and the subcarrier is then suppressed so that only the Q and I sidebands remain.

The composition of a color signal is shown in Figure 13-5. The Y signal contains information for black-and-white receivers as well as for the color receiver; since its bandwidth is 4.2 MHz, the detail from this signal on a black-and-white receiver is somewhat better than that from a normal black-and-white signal. The Q and I signals are at the same 3.58 MHz frequency, but they are 90 degrees out of phase with each other. The components of the signal are interleaved by the transmitter into clusters of energy so they do not interfere with one another.

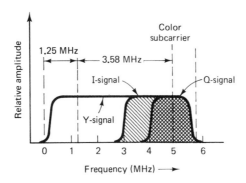

Figure 13-5. Color TV Signal Components

Figure 13-6 shows a composite video signal for a black-and-white receiver. In a single line of video on an American TV set, the receiver scans the picture tube 525 times, producing a picture. A horizontal blanking pulse cuts off the screen during retrace. Since the U.S. standard is 30 frames per second, the picture tube is scanned 15,750 times per second. Thus, horizontal sync pulse frequency is 15,750 Hz.

Vertical synchronizing pulses at 60 Hz and horizontal sync pulses at 15,750 Hz maintain synchronization between transmitter and receiver. Horizontal blanking pulses turn off the picture tube as the trace snaps back across the screen to start a new line, while the vertical blanking pulses turn off the screen while the trace is returning from the bottom to the top of the screen.

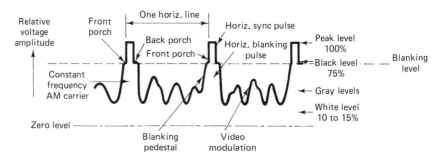

Figure 13-6. Composite Video Signal

The portion of signal between the horizontal blanking and horizontal sync pulses is the pedestal, and is broken into the "front porch" and the "back porch." In a color transmission, as shown in Figure 13-7, the back porch of the horizontal blanking pulse is used to carry color information; the 3.58 MHz color burst subcarrier sits there.

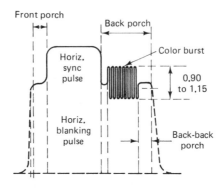

Figure 13-7. Color Burst Subcarrier on the Back Porch

The block diagram of a typical solid-state color TV system is shown in Figure 13-8. In the tuner section, the selected VHF or UHF signal is mixed with a local oscillator signal in order to generate signals that are modulated by sound and video. The FM sound signal is applied to sound amplifying and detecting circuitry, where it is amplified, detected, and converted to AF in the same way as in a normal FM receiver.

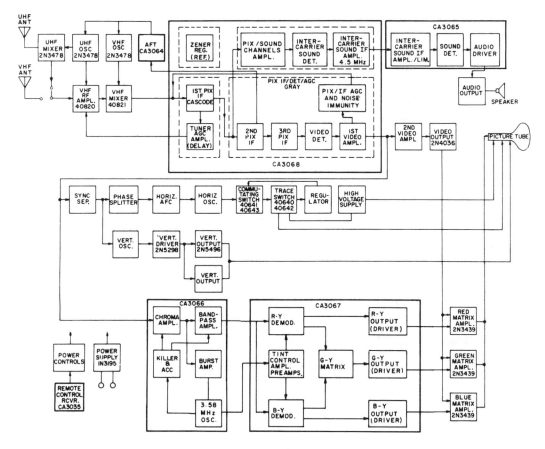

Figure 13-8. Color TV Block Diagram

The AM signal containing video is amplified, then the video is detected, amplified further, and applied to the respective video amplifiers for red, green and blue. These outputs are then applied to the grids for these colors in the picture tube. (For a black-and-white signal, only one grid is used.)

For sweep and retrace timing, the sync separator peels off the horizontal and vertical synchronization pulses from the video. The horizontal sync pulses at 15734.264 Hz trigger the horizontal oscillator at the appropriate times. This signal is converted to a sawtooth and applied to the horizontal deflection circuitry. Horizontal AFC helps give the oscillator noise immunity and maintains it at the proper frequency.

In the vertical oscillator and driver, the 60 Hz (actually, 59-94 Hz) vertical sync pulse is also converted into a sawtooth waveform, then converted into a trapezoidal signal for application to the vertical sweep transformer and deflection yoke windings.

To this point, a color and a black-and-white receiver are basically the same. However, the color set also has a *chrominance* channel, which amplifies the chrominance signal, detects the individual red, green and blue components, and sends them to the picture tube drivers. Since the color carrier is suppressed at the transmitter, a 3.58 MHz oscillator reinserts this carrier so the color signals can be demodulated. Automatic color control (ACC) controls the amplitude of the color bursts much as AGC controls gain, and the "color killer" blanks the color circuitry when only a black-and-white signal is being received. This prevents colored "noise" from appearing on the screen.

The TV set outlined above is built for the NTSC color system used in the U.S. and Japan. In Europe, where the PAL (phase-alternation line system) is used, the phase of the color subcarrier is changed from scanning line to scanning line, requiring the transmission and processing of a line switching signal as well as a color burst.

This chapter contains many radio and television receiver circuits. (*Figure 13-8 courtesy of RCA Solid State.*)

AM/FM AUTO RADIO WITH MONO AUDIO OUTPUT

The circuit of the Automatic 6800 and 7C series AM/FM auto radios is given in Figure 13-9. The mode switch (SW-1), when set in the AM position as shown in the diagram, feeds the audio output at the AM detector diode D1 through a low-pass filter to pin 14 and out pin 12 of IC-1 on the FM IF circuit board at the lower right of the diagram. SW-1 also connects the positive DC voltage buss of the AM tuner to the A+ terminal of the FM IF board, which is connected through R6, L1 and on-off switch SW-2 on the audio amp board, and then through L5 to the positive side of the vehicle battery. L1 utilizes IC-1 on the AM ASSY board as the RF amplifier, mixer and local oscillator; these are tuned by permeability-tuned coils L2, L3 and L4. This IC also contains the IF amplifier which utilizes IF transformers T1 and T2 for providing selectivity.

The FM tuner at the lower left also uses permeability-tuned coils for tuning through the 88-108-MHz FM band. L1 is the RF amplifier input coil, L2 is the mixer input coil and L5 is the local oscillator coil. The RF amplifier employs FET Q1 and the LO and mixer employ NPN transistors Q2 and Q3.

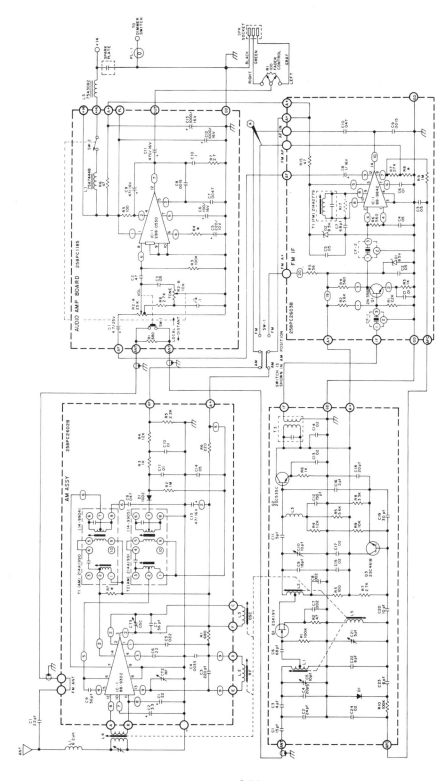

Figure 13-9. AM/FM Auto Radio with Mono Audio Output

The 10.7-MHz IF signal is fed from the secondary of IF transformer T3 on the FM tuner board to the base of IF amplifier Q1 on the FM IF board through a 10.7-MHz selectivity filter. The output of Q1 is fed through another selectivity filter to the input of IC-1. This IC contains IF amplifiers, limiters and the quadrature FM detector whose phase-shift resonant circuit is connected to pins 9 and 10 of the IC. The coil T1 includes the shunt capacitor. The core is adjusted for maximum receiver audio. When the recovered audio is distorted due to clipping, R17 is shunted across T1 to reduce the Q of the coil. The speaker socket enables connection of a front speaker and a rear speaker and the fader (R1) can be adjusted to control the relative volume of one speaker with respect to the other. (*Courtesy of Automatic Radio.*)

AM RADIO TUNER USING AN IC

The LM3820 IC is used in the AM broadcast band tuner circuit given in Figure 13-10. The input signal is picked up by the ferrite rod antenna or

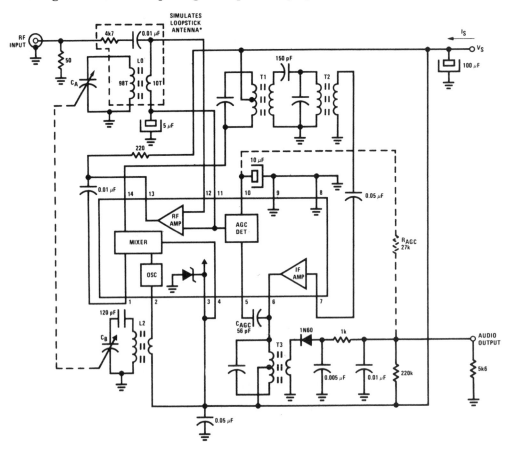

Figure 13-10. AM Radio Tuner Using an IC

an external antenna. The loop antenna and the oscillator coil (L2) are tuned by a two-gang tuning capacitor with its 140-pF section connected to the loop antenna and its 60-pF section connected across the oscillator coil. The 14-pin dual in-line IC can be operated from a DC source in the 12-24 volt range. The typical sensitivity is 35 microvolts at the antenna terminal for 10 millivolts of audio output. The IF transformers T1, T2 and T3 are tuned to 455 kHz, the IF signal frequency. (*Copyright National Semiconductor Corporation.*)

AUTOMATIC FREQUENCY CONTROL

Automatic frequency control is used in many FM receivers to compensate for local oscillator frequency drift, and it was used in the 1930's in some AM broadcast band receivers. It is used in many TV sets, where it is called *automatic fine tuning*. An IC designed for use in TV AFT applications can also be used in radio receivers. The Sprague ULC-2264 IC can be used in the circuit given in Figure 13-11 to provide AFC in radio receivers. The IF signal is monitored and any frequency error causes an error voltage to be applied to a varactor diode, which tunes the oscillator until the frequency error is zero. The frequency error voltage appears at pins 5 and 8. When there is no frequency error the voltages at pins 5 and 8 are equal. But when the frequency error is upward or downward from the IF reference frequency, the error voltage at pins 5 and 8 differ and apply the correct amount of voltage to the varactor to increase or decrease its capacitance to correct the frequency error. (*Courtesy of Sprague Electric Company.*)

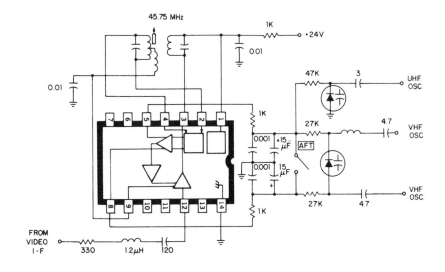

Figure 13-11. Automatic Frequency Control

AUTO RADIO/TAPE PLAYER

The circuit of the Sparkomatic SR46 AM/FM auto radio with a built-in 8-track tape deck is given in Figure 13-12. The FM front end, shown here only as a block, enables tuning the receiver through the 88-108 MHz FM broadcast bands. The 10.7-Hz IF signal is fed through a selectivity filter to the input of IC101, whose output in turn is fed through a second selectivity filter to the input of C102, which contains IF amplifier and limiter circuits plus a quadrature FM demodulator. The recovered audio is fed through volume control VR102 to Q102 which is the automatic switching transistor. Its output is fed to IC103, the PLL stereo decoder. The L+R signals and the L−R signals are fed through a matrix which feeds the left signal to the left channel of the audio system and the R signals to the right channel. When set in the AM mode, monaural sound is amplified and reproduced by the speakers. When playing 8-track tape cartridges, the outputs of the tape heads are fed to the left and right channel amplifiers. (*Courtesy of Sparkomatic Corporation.*)

CHROMA SYSTEM

Figure 13-13 illustrates a chroma subcarrier regeneration system for a color television receiver. The 3.58 MHz oscillator is controlled by a crystal attached to pins 6 and 7 of the ULN-2124A IC. This oscillator provides the necessary signal for reinsertion of the color subcarrier into the color sidebands. The oscillator is stabilized by the automatic phase control (APC) circuitry.

The two oscillator outputs, from pins 2 and 3 of the ULN-2124A, are applied to a quadrature network, which results in the inputs to the demodulator (pins 7 and 8 of the ULN-2269A IC) being 90 degrees out of phase with each other.

One color demodulator processes the color signal and an oscillator signal to produce a composite B-Y (that is, blue-Y, where Y is the black-and-white signal) output. Another demodulator produces an R-Y (red and Y) output. A matrix circuit combines B-Y and R-Y to produce the G-Y (green and Y) output. The circuit has an adjustment for the color killer threshold, as well as adjustments for automatic phase control, automatic color control, and hue. (*Courtesy of Sprague Electric Company.*)

CRYSTAL SET

The crystal set was a popular radio receiver in the 1920's and early 1930's. It was not only used for listening to radio broadcasts, but it was used on many ships as the primary radio receiver. In spite of its short range and lack of sensitivity, it was used on ships because it required no electric power supply and was extremely simple. Thousands of crystal sets were bought for home use. One of the major suppliers was Philmore, which was based in the Borough of Queens in New York City.

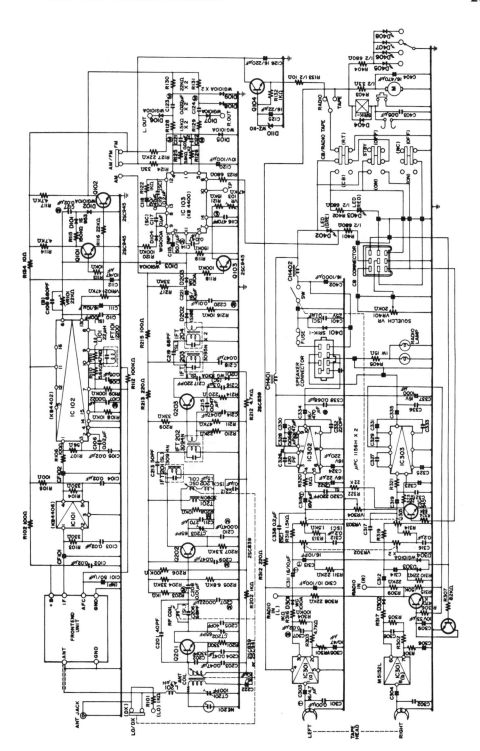

Figure 13-12. Auto Radio/Tape Player

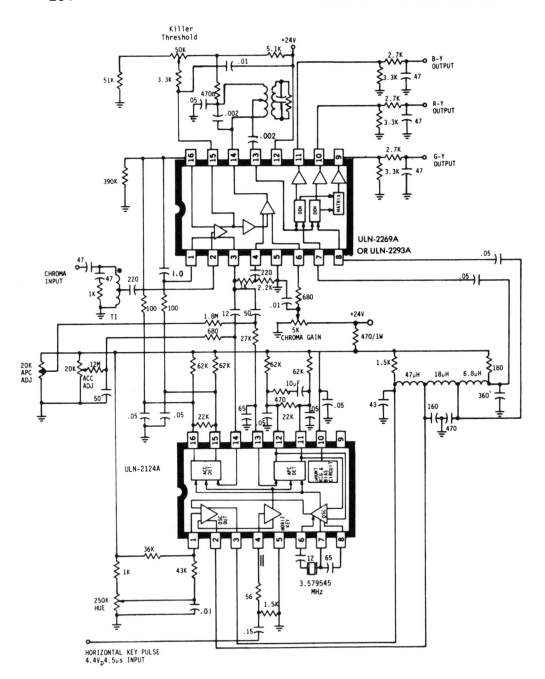

Figure 13-13. Chroma System

The simplest of all radio receivers consists of a razor blade detector connected to a wire antenna and an earth ground. A pair of headphones is connected across the detector. The razor blade works like a crystal. When the blade is connected to ground and a catwhisker connected to the antenna touches the edge of the blade, the poor electrical contact between the catwhisker and the blade forms a rectifier which erases one-half of the modulation envelope and also enables the audio modulation to be recovered from the radio signal.

The same circuit can be used with a galena or silicon crystal. The crystal is a better rectifier than the razor blade. The catwhisker is touched to the surface of the crystal. Not all parts of the crystal surface are satisfactory for rectification. You must try various parts of the crystal surface until you find one that produces the loudest signal in the headset. The catwhisker is a part of the crystal detector assembly shown in Figure 13-14(a).

A schematic of the same receiver is given in Figure 13-14(b). Diode D represents the crystal detector. You can use a germanium or silicon diode in lieu of the crystal detector. A diode does not require any adjustment, but it won't be as sensitive to weak signals because of the surface barrier voltage of the diode. When a germanium diode (such as a 1N34) is used, the receiver sensitivity is about 0.2 volt. Below that voltage, the diode will not conduct and no audio signal will be heard. A silicon diode (such as a 1N60) will not

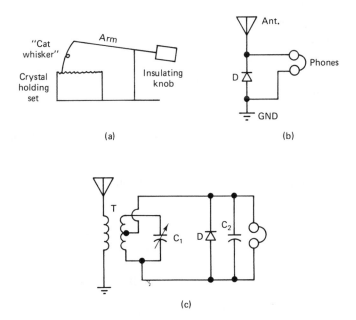

Figure 13-14. Crystal Set

conduct until the applied signal has a level of 0.6 volt or higher. But for receiving powerful local AM broadcast stations, either a germanium or silicon diode should work very well.

This crystal set has very poor selectivity since it has no tuned circuit. It is broadly resonant at the resonant frequency of the antenna system. Greater selectivity can be obtained by adding an RF transformer (T), a variable capacitor (C1) and a fixed capacitor (C2) which is connected across the headphones and to ground which bypasses any RF remaining after the diode. But, the secondary of T which is tuned to the frequency of the intercepted station by adjusting C1 is loaded down by the low impedance of the crystal diode detector circuit.

This can be corrected by using a loopstick antenna in lieu of an RF transformer as shown in Figure 13-14(c). The loopstick should have a tapped main winding and a low impedance winding to which the antenna and ground are connected. The tap of the main winding is the detector signal take-off point. This circuit is quite selective but requires an outdoor wire antenna (about 50 feet long) plus a good earth ground. The ground is essential.

DISCRETE CRYSTAL CB TRANSCEIVER

The circuit shown in Figure 13-15 is of the receiver of a CB transceiver that can be operated on up to 12 different channels. A separate transmit crystal and a receive crystal are required for each channel. A ganged channel selector switch selects the frequency-determining crystals.

In the receiver circuit, the incoming signal from the antenna is fed through the T/R relay (not shown) and transformer T11 to the RF amplifier stage, Q4. The receiver frequency is determined by the Y2 crystal that is selected by switch S1B. Output from this receive crystal is mixed with the RF signal in the first converter Q5, and the resulting intermediate frequency signal is then coupled to T4, the first IF transformer. If the transceiver is set to 26.965 MHz, for example, the receiver crystal operates at 1651.5 KHz above or below 26.965 MHz.

The second converter does not use a crystal-controlled oscillator. Instead, the second LO coil (L5) is tuned to 2106.5 kHz so that a 455 kHz second IF is generated. Detection is performed by CR3 and associated circuitry.

Potentiometer R36, by controlling the bias on Q10, provides squelch control; when the incoming signal is strong enough, it causes Q10 to cut off, allowing Q11, the first audio stage, to turn on and amplify the audio signal. (In transmit mode-10V is applied to the wiper arm of R36, and through R41 to the base of Q10.)

After amplification by Q11 and Q12, the audio signal is transformer-coupled to the push-pull output stage, consisting of Q13 and Q14. (*Courtesy of Hallicrafters.*)

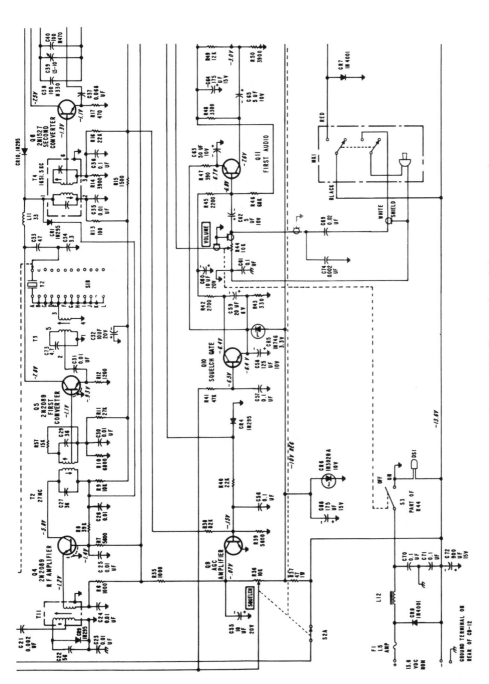

Figure 13-15. Discrete Crystal CB Transceiver (also on next page)

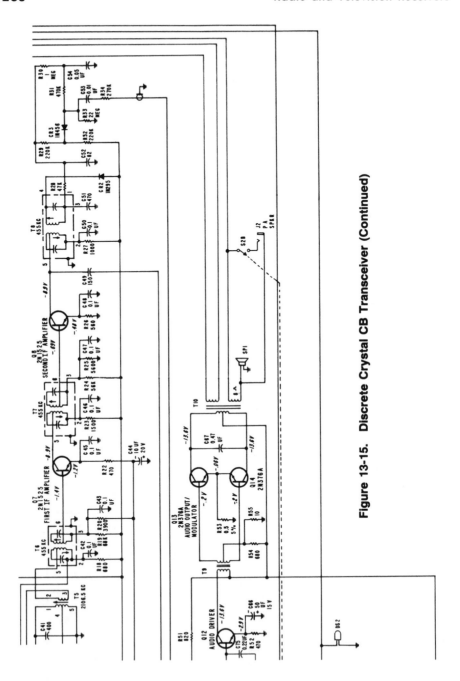

Figure 13-15.   Discrete Crystal CB Transceiver (Continued)

FLEWELLING RECEIVER

Figure 13-16 is a solid-state version of the once-popular receiver circuit devised by E. T. Flewelling which uses capacitive coupling to the antenna and a switch to convert the receiver from a regenerative circuit to a super-regenerative circuit. When switch S is set to the position where the capacitor between the source of the FET and the tank coil is shorted out, the circuit is regenerative; when S is in the other position, the circuit squeggs and functions as a super-regenerative detector. Regeneration is controlled by rotating the tickler coil to vary the coupling between the tank and tickler coils.

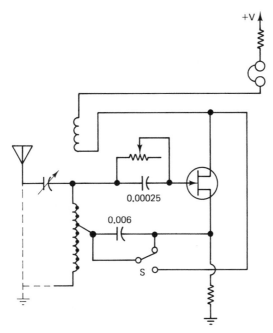

Figure 13-16. Flewelling Receiver

NOISE-OPERATED SQUELCH

Most FM communications receivers use a noise-operated squelch to silence the speaker between intercepted transmissions. The noise-operated squelch is said to have been invented by the late Dr. Daniel Noble who was executive vice president of Motorola. This squelch circuit senses both the level of the intercepted radio signal by monitoring the limiter voltage and the quieting of the noise at the detector output and intercepted radio signal. Figure 13-17 is a schematic of the noise-operated squelch circuit used in the Quintron RC-151 and RC-451 receivers that are designed for use at base and repeater stations.

The noise from the detector output is fed through connector J1103 from where it is routed through a high-pass filter (C1100, C1101, R1102 and R1104) to the base of the active filter transistor (Q1100). The noise is fed to the input of IC1100A where it is amplified and filtered. The output of IC1100 is fed to the noise rectifier (CR1100, CR1101) which converts the noise into a

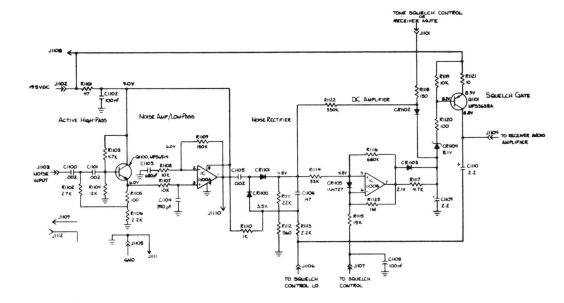

Figure 13-17. Noise-Operated Squelch

DC signal which is amplified by IC1106. The DC is fed through a Zener diode (CR1104) to the base of squelch gate transistor Q1101. A tone squelch control can be connected to the noise squelch circuit through connector J1101. The Zener diode will avalanche when the potential across it reaches 5.1 volts. When the DC output of IC1106 is less than 5.1 volts, the Zener keeps the squelch from muting the receiver.

When no signal is being received, self-generated receiver noise at the detector output mutes the speaker. When a signal is intercepted, the noise quiets and the squelch awakens and allows the radio signal to be heard. The effective sensitivity can be controlled with the squelch control which is connected through connectors J1106 and L1107. (*Courtesy of Quintron Corporation.*)

ONE-FET RADIO RECEIVER

A single FET can be used in the circuit given in Figure 13-18 to form a sensitive and relatively selective radio receiver. It can be used with a pair of 2000-ohm headphones or as a tuner with a hi-fi amplifier. The receiver is tuned by adjusting C1 which is connected across the secondary of RF transformer T. Because the input resistance of an FET is extremely high, the tuned input is not loaded down significantly by the FET and its selectivity will not be impaired. The FET is wired into a power detector circuit which is similar to a plate detector circuit using a triode tube. The FET gate is biased negatively by the voltage drop across R1, the source bias resistor. The headphones are connected in series with the drain circuit of the FET. By connecting the headphones between the negative terminal of the 9-volt transistor battery and ground, one side of the headphone will be at ground potential.

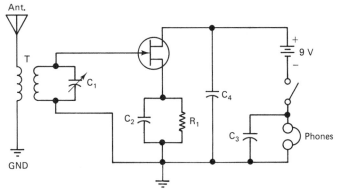

Figure 13-18. One-FET Radio Receiver

PRESSLEY IMPEDANCE BRIDGE SUPERHETERODYNE

An electrical bridge is used in the superheterodyne receiver circuit as devised by Jackson H. Pressley, chief radio engineer of the U.S. Signal Corps during the 1920's. A solid-state circuit of a receiver front end using an impedance bridge is given in Figure 13-19. Only one field-effect transistor is required to serve the functions of the mixer and local oscillator. When the Wheatstone bridge is balanced, the local oscillator is tuned with C3 and the mixer input is tuned with C4 without one affecting the other, and LO radiation is minimized. C3 is used for tuning in stations and C4 is adjusted to obtain maximum signal level. The IF signal is selected by T2 which feeds it to the IF amplifier (not shown). In the tube version, which was discussed in the February 8, 1925, edition of the *New York Herald Tribune Radio Magazine*, Pressley used a pair of 150-pF capacitors for C1 and C2.

REINARTZ RECEIVER

A solid-state version of the Reinartz receiver is shown in Figure 13-20. In this circuit, which was invented by John L. Reinartz, a coil in the drain circuit of the FET is used to provide regenerative (or positive) feedback to the antenna coil. The incoming signal is tuned by capacitor C1 and coil L2, rectified by diode D, and amplified by Q. Coil L3 is the tickler coil, and capacitor C3 is used to adjust the amount of regenerative feedback to give the circuit maximum gain without going into oscillation.

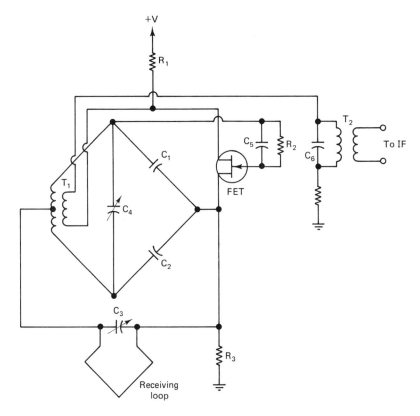

Figure 13-19. Pressley Impedance Bridge Superheterodyne

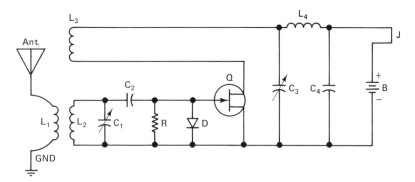

Figure 13-20. Reinartz Receiver

SUPERREGENERATIVE VHF RECEIVER

One transistor and an IC are used in the 2-meter band super-regenerative receiver circuit given in Figure 13-21. The antenna may be a 19-inch vertical whip or stiff wire that is capacitively coupled to one end of the tank

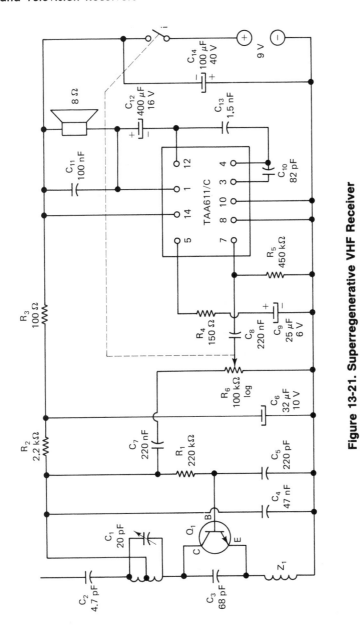

Figure 13-21. Superregenerative VHF Receiver

circuit. The audio output is taken from a tap on the tank coil. Feedback is provided by C3 which is bridged across the collector-emitter path of Q1. The frequency range can be selected by using coils of the proper inductance for the tank and for the RF choke (Z1) in the emitter circuit. The audio output is fed through C7 and volume control R6 to the input of the TAA611/C IC. Almost any AF amplifier IC can be used in the circuit.

TV HORIZONTAL SWEEP

The horizontal sweep circuit in Figure 13-22 uses a single IC for horizontal processing. The sync signal is applied through pin 3 to a phase detector within the IC. Output from the phase detector is coupled to pin 7, where it controls the timing of an internal oscillator whose output, in turn, is taken off pin 1 and applied to the horizontal driver.

The square wave applied to the base of the first transistor is converted to the requisite sawtooth by the hysterisis curve of the horizontal drive transformer and its associated circuitry. This sawtooth is applied through the horizontal output transformer to the picture tube yoke, causing the trace to sweep across the screen at 15,750 Hz. (*Courtesy of Sprague Electric Company.*)

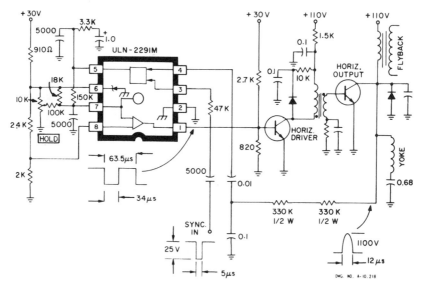

Figure 13-22. TV Horizontal Sweep

TV IF AMPLIFIER AND DETECTOR

Figure 13-23 shows a practical IF amplifier and detector for a television receiver. The circuit uses an MC1350 integrated circuit as an IF amplifier and an MC1330 IC as the detector. This circuit has a typical voltage gain of 84 dB and an AGC range of 80 dB. For a 45 MHz IF, C10 is 33 pF and L3 is 10 turns of #26 wire on a 3"-long, 3/16" core. Fine tuning is achieved by distorting the coils of L3. (*Courtesy of Motorola, Inc.*)

TV SOUND DETECTOR

Figure 13-24 shows a detector using a single ULN-2212B integrated circuit to provide one watt of audio output into 8 ohms for a television sound

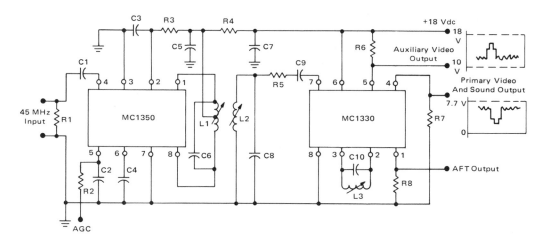

Figure 13-23. TV IF Amplifier and Detector

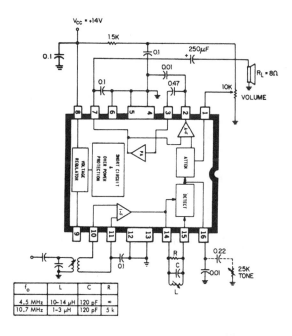

Figure 13-24. TV Sound Detector

channel or FM radio AF detector/amplifier. The IC employs a quadrature detector, and the values of the phase-shifting network across pins 14 and 15 can be altered for either the 4.5 MHz IF found in a TV set or the 10.7 MHz IF found in an FM radio. Attached to pin 16 is a de-emphasis network. (*Courtesy of Sprague Electric Company.*)

UHF RECEIVER PRESELECTOR

Some UHF communications receivers are equipped with a front end preselector like the one shown in Figure 13-25. It consists of five resonant cavities which are tuned to pass the required band of frequencies. It also has two additional resonators which are tuned to the injection frequency for the mixer. These resonant cavities eliminate harmonics and other spurious signals. The preselector has three interfaces: a jack for the RF input signal, a shielded cable with a phono plug that connects to the IF amplifier input and another shielded cable, also with a phono plug, that connects to the local oscillator frequency multiplier. (*Courtesy of Aerotron, Inc.*)

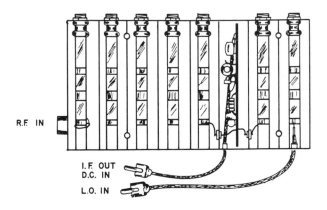

Figure 13-25. UHF Receiver Preselector

ULTRA AUDION RECEIVER

The circuit of a solid-state version of the De Forest ultra audion regenerative receiver is given in Figure 13-26. The feedback path is through the drain-gate capacitance. The tuning capacitor (C2) is in series with the tank coil (L1). Regeneration is controlled with the potentiometer (R2) which is used to vary the drain voltage applied to the FET.

VHF RECEIVER FRONT END UTILIZING FOUR MOSFETs

The use of four single-gate 3N128 MOSFETs in the RF amplifier, the mixer, IF preamplifier and local oscillator gates of a VHF receiver front end is shown in Figure 13-27. The 50-ohm input is fed to a tap on the input autotransformer which is tuned with a 1-9 pF variable capacitor. The output of the RF amplifier (Q1) is capacitively coupled to a tap on the mixer input autotransformer. The RF amplifier is neutralized by an 0.5-3 pF trimmer capacitor. The LO (Q4) output is loosely coupled to the mixer (Q2) input coil. The 30-MHz IF signal is coupled through an IF transformer to the gate of

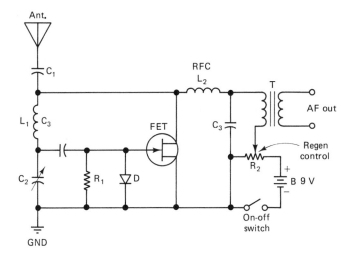

Figure 13-26. Ultra Audion Receiver

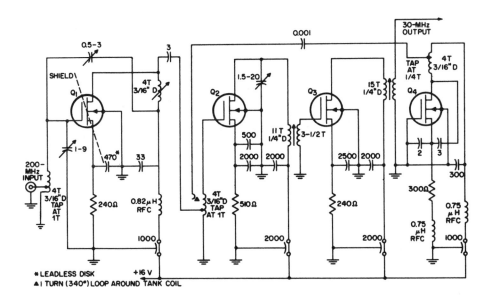

Figure 13-27. VHF Receiver Front End Utilizing Four MOSFETs

the IF preamplifier (Q3) whose 30-MHz output is fed through another IF transformer to the main IF amplifier (not shown). The values and the coil turns noted on the diagram are for receiving on 200 MHz. The component values and the number of RF amplifier, mixer input and LO coil turns can be changed to permit operation at another frequency. (*Courtesy of RCA Solid State.*)

VIDEO SIGNAL PROCESSOR

The circuit shown in Figure 13-28 utilizes a single ULN-2125A integrated circuit to perform much of the processing of a black-and-white or color TV signal. Video into pin 8 is fed through an emitter-follower and is available as low-impedance video at pin 9. High-impedance, noise-canceled video is produced at pin 5. After the video is stripped and the signal processed by an internal sync separator, both positive and negative sync pulses are produced. The circuit also provides forward AGC for the video IF stages, delayed forward AGC for an NPN bipolar tuner, and delayed reverse AGC for a PNP, MOSFET, or tube-type tuner. (*Courtesy of Sprague Electric Company.*)

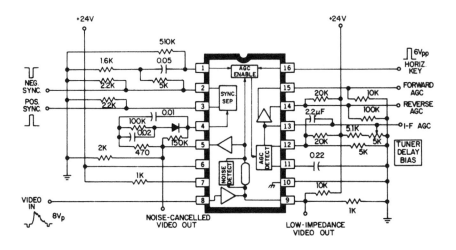

Figure 13-28. Video Signal Processor

VOLTAGE DOUBLER DIODE RADIO RECEIVER

A pair of germanium diodes can be connected in a half-wave voltage doubler detector circuit as shown in Figure 13-29. This detector has a voltage gain of 6 dB over a conventional single diode detector. When the input signal at the tap on L2 is swinging positive, diode D1 conducts and allows the received audio signal to be developed across C1. When the signal polarity reverses, shunt diode D2 conducts and causes C2 to become charged. The charges in C1 and C2 are in series-aiding and will be equal to double the charge in either C1 or C2 alone. By the same token, the level of the recovered audio signal will be equal to the sum of the audio voltages across C1 and C2.

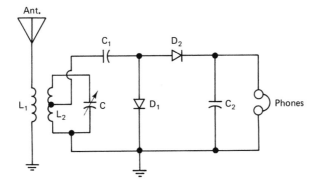

Figure 13-29. Voltage Doubler Diode Radio Receiver

CHAPTER 14

Radio Transmitters

INTRODUCTION

A radio transmitter is a generator of radio frequency energy. It is really nothing more than an oscillator followed by several stages of power amplification. This amplification is designed to boost the oscillator signal up to the desired power level, be it 500 milliwatts or 50,000 watts. In addition, the transmitter usually contains a modulator so that the RF energy can be used for conveying intelligence such as voice, music, data, etc.

In the early days of radio, the first transmitters were of the damped-wave type, using a spark gap or other means of producing RF energy in bursts. Their use is no longer authorized, since they cause severe interference to other radio systems. Unmodulated CW transmitters supplanted the spark gaps, and these kinds of transmitters are still in use, especially for amateur radio and long-distance communication. In a CW transmitter, a telegraph key or other switch is used to interrupt the transmission of a signal. CW transmitters are also used for generating heat in industrial and medical applications.

An amplitude-modulated (AM) transmitter, like a CW transmitter, consists basically of an oscillator followed by several stages of amplification and isolation. However, a modulator is also added that causes the output power of the transmitter to vary at the modulation rate, as shown in Figure 14-1. When the RF amplifier is modulated 100 percent, the output power from the transmitter rises 50 percent above the unmodulated carrier signal level.

A double sideband (DSB) transmitter is a type of AM transmitter which uses a balanced modulator. This modulator cancels out the carrier frequency but continues to generate the two modulating sidebands. For single sideband transmission, a filter is used to cancel out one of the sidebands, leaving only the upper sideband (USB) or the lower sideband (LSB) to carry the modulating information. All of the available power is contained in the transmitted sideband, making this type of transmission very efficient. Virtually no signal is transmitted unless modulation is applied.

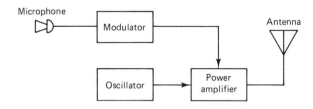

Figure 14-1. Basic AM Transmitter

In a frequency-modulated (FM) transmitter, intelligence is imparted by varying the oscillator frequency, or, as shown in Figure 14-2, by varying the phase of the transmitted signal. The output power of an FM transmitter remains constant, but its frequency varies with modulation.

This chapter illustrates and describes discrete transmitters as well as the transmitter sections of several transceivers.

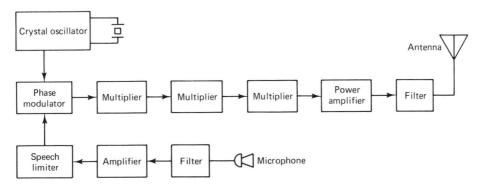

Figure 14-2. Phase-Modulated Transmitter

AMPLITUDE MODULATION LIMITER

The modulation limiter circuit used in an AM CB walkie-talkie whose circuit is given in Figure 14-3 senses the level of the audio signal at the output of the modulator. As can be seen in the diagram, the modulated emitter voltage line to the final RF power amplifier (TR16) is connected to the modulation limiter input through C44 and R53. The AF signal at the junction of R53 and modulation level control R52 is rectified by diode V2. Its positive DC output is filtered by C43 and C42 and is fed through R57 to the base of TR11. The TR11 collector voltage is obtained from the emitter of AF amplifier TR9. When the AF voltage at the input of the modulation limiter rises, the TR11 collector current increases. This causes the voltage drop across the TR9 emitter resistor (R40) to rise also. This causes the TR9 emitter bias to rise. This emitter bias is in opposition to the fixed forward bias on TR9 and results in a reduction of TR9 gain due to the reduction in forward bias. (*Courtesy of SBE Inc.*)

AM WALKIE-TALKIE

A schematic of a simple AM walkie-talkie is given in Figure 14-4. It utilizes a single PNP transistor (TR-1) as a crystal-controlled autodyne mixer which also functions as the local oscillator. The 455-kHz IF signal is fed through IF transformer T-1 to the base of the IF amplifier transistor (TR-2) whose output is coupled through IF transformer T-2 to diode CR-1. The detector output is fed through volume control R8 and the PTT switch to the base of AF amplifier transistor TR-3. Its output is fed through AF transformer T-3 to the base of power transistor TR-4 which drives the 100 ohm speaker-microphone through output transformer T-4.

When the PTT switch is pressed in, power is applied through one of the windings of T-4 to the transmitter oscillator transistor TR-5 which also functions as the final RF output stage of the transmitter. The output of the

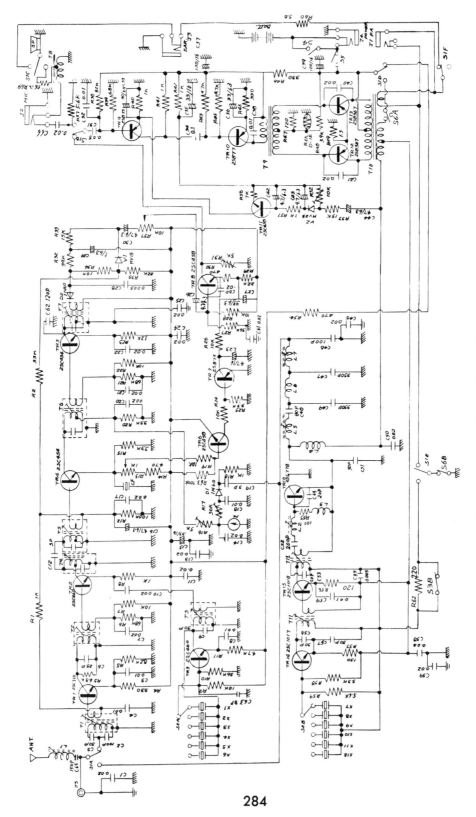

Figure 14-3. Amplitude Modulation Limiter

284

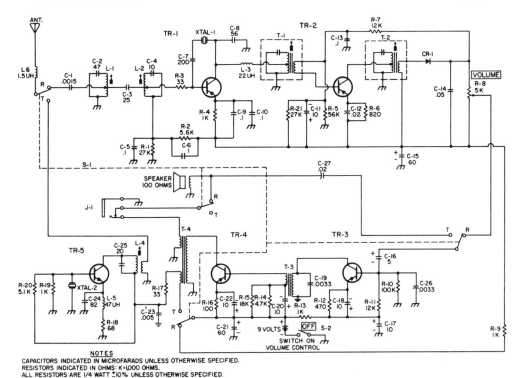

Figure 14-4. AM Walkie-Talkie

speaker microphone is fed to the base of TR-3 which amplifies the voice signal and feeds it to TR-4 which now serves as the amplitude modulator. The audio developed across T-4 alternately bucks and boosts the instantaneous DC voltage applied to the collector of TR-5, causing the output power to vary up and down. This walkie-talkie was designed for use in the 27-MHz citizens' band without a license. Since unlicensed operation, under Part 15, FCC Rules and Regulations, is banned as of 1983, and because new transmitters may not be used in the 27-MHz band without a license, this walkie-talkie circuit could be used in a 49.82-49.90 MHz band only when the total input power is less than 100 milliwatts and antenna length is less than 1 meter. It can be used without a license if certified as complying with the applicable Part 15 rules. (*Courtesy of CB Magazine.*)

EMERGENCY TRANSMITTER

The diagram given in Figure 14-5 is of the Microlert transmitter, which was designed for use by a handicapped and ill person for summoning medical aid when required. It also can be used for transmitting a security alarm. This tiny transmitter, smaller than a pocket watch, utilizes a single transistor in a Hartley oscillator circuit. It is digitally modulated by IC U1. The trans-

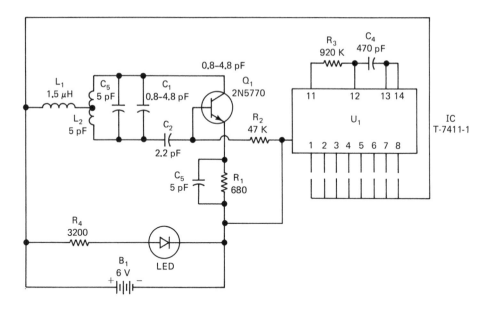

Figure 14-5. Emergency Transmitter

mitter is turned on by squeezing the contacts that touch the positive and negative ends of the tiny 6-volt battery. This transmitter utilizes its less-than-a-single-turn tank coil (L2) as its antenna. It can be adjusted to operate in the 300-MHz range, as permitted by FCC Rules, Part 15. When keyed on, it is intercepted by a wall-mounted receiver which responds only to the digital code signal emitted by the transmitter. When its signal is intercepted by its associated receiver, an automatic telephone dialer dials the phone number of the person who is being depended upon for assistance.

FM TRANSMITTER EXCITER

An FM transmitter exciter generally consists of a modulation limiter, an oscillator, modulator and one or more frequency multiplier stages. A typical FM exciter circuit is shown in Figure 14-6. The audio input signal from a microphone is fed in at J101 and J102. It is fed through a 1000-ohm potentiometer, through an RC filter (C01, C02, C04, R02, R03 and R04), to the base of transistor Q01, which with Q02 forms an amplitude limiter. The signal developed across R08, the emitter resistor shared by Q01 and Q02, is the input signal for Q02, since its base is grounded for AF through C05. The output of Q2 is direct-coupled to the base of Q03, an emitter-follower whose output signal is fed to the hard limiter comprising diodes CR01 and CR02. These diodes are normally forward-biased by the positive voltage fed to their anodes through R13.

The limited audio signal is then fed through active filters Q04 and Q05. The output of Q05 is direct-coupled to the base of Q06, which is a PNP

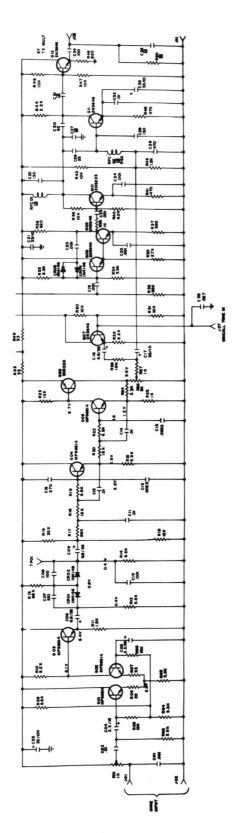

Figure 14-6. FM Transmitter Exciter

transistor whose emitter is connected to the V_{DD} bus and whose collector load resistance is potentiometer R06, the FM deviation control. The signal is then fed to the emitter of Q07 through C18, and through C17 and RFC02 to the base of Q11. The output of the tone squelch encoder is fed through J107 to the base of Q07, where the voice and tone modulation are mixed.

The RF signal selected from among four transmit oscillators by the channel selector switch is fed through C19 to the base of Q08, an emitter follower that is direct-coupled to the base of Q09, whose output in turn is fed through R38 and C23 to the base of Q10. The RF signal at the collector of Q10 is fed through C29 to the base of phase modulator transistor Q11 and also through C29 to the collector of Q11. As stated earlier, the processed audio modulating signal (voice and sub-audible tone) is fed to the base of Q11. When the Q11 collector current is varied by both the RF signal and the AF signal, the phase of the output signal will vary, giving the effect of narrow band frequency modulation. The resulting equivalent of an FM signal is fed through C31 to the base of Q12, which serves as a frequency multiplier. (*Courtesy of Aerotron, Inc.*)

FM WIRELESS MICROPHONE

The FM wireless microphone whose circuit is given in Figure 14-7 is operable within the 88-108 MHz FM broadcast band. It can be tuned by adjusting the core of L1. The frequency of the RF signal is deviated by the variation in capacitance of transistor Q1. This capacitance changes due to the AF signal picked up by the microphone, which may be of the crystal, ceramic or dynamic type. C1 and C2 are 10-pf disc capacitors, C3 and C4 are 470 pF, C5 and C6 are 10-uF electrolytics. R1 and R3 are 100,000-ohm resistors, R2 is a 10,000-ohm resistor and R4 is a 470-ohm resistor. L1 may be

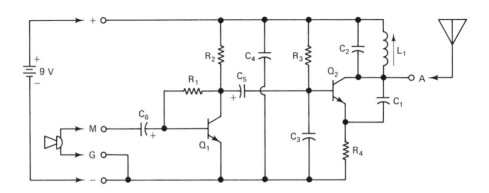

Figure 14-7. FM Wireless Microphone

an inductor whose inductance can be adjusted through the 0.2-0.3 microhenry range. (*Courtesy of Ramsey Electronics Inc.*)

GMRS FM TRANSMITTER

The circuit of a UHF/FM transmitter designed especially for General Mobile Radio Service applications is given in Figure 14-8(a). The oscillator (Q301) operates at 1/18th of the carrier frequency. When a 26-MHz crystal is used, the transmitter will operate at 468 MHz. The oscillator output is fed through a buffer stage to the first frequency tripler (Q307) which in turn feeds its output to frequency doubler Q308. The signal is then fed through two amplifier stages to the second tripler (Q311) which drives the RF driver amplifier (Q312) which delivers its signal to the RF power amplifier through a cavity filter.

The RF power amplifier circuit is shown separately in Figure 14-8(b). This RF amplifier delivers 5 watts of RF output into a 50-ohm antenna system. The microphone output is fed to the input of IC Q371 whose output is fed through a roll-off filter and modulation deviation control (R381) to varactor diodes (Q304 and Q303) which are across the signal path between the oscillator and the buffer amplifier. These varactors phase modulate the oscillator signal. The FM deviation introduced by the modulator is multiplied 18 times. The output signal can be set with R381 to maximum FM deviation of ±5 kHz. (*Courtesy of Standard Communications Corporation.*)

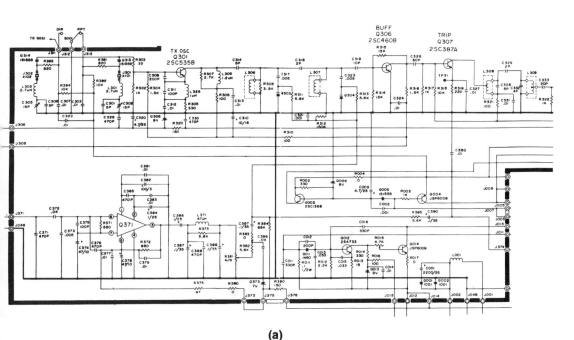

(a)

Figure 14-8. GMRS FM Transmitter

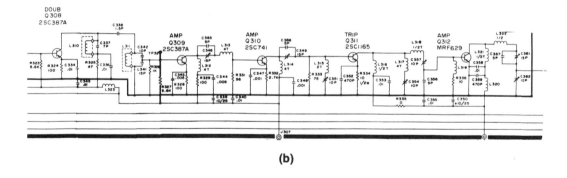

(b)

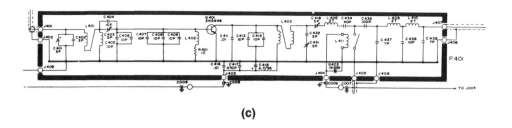

(c)

Figure 14-8. (Continued)

LICENSE-FREE AM TRANSMITTER

The circuit of a 49-MHz band AM transmitter is given in Figure 14-9. It utilizes a quartz crystal (Y) to stabilize the oscillator frequency. The audio input device is a PM speaker that is used as a microphone. It is coupled to the gate of Q1 through T1, an 8:50,000-ohm step-up transformer which provides around 38 dB of voltage gain. The drain of Q1 is coupled through C2 to the gate of Q2, the modulator. The modulation transformer (T2) causes the audio output voltage to alternately buck and boost the instantaneous drain voltage at Q3. The total input power must be limited to 100 milliwatts as specified in Part 15, FCC Rules and Regulations. The length of the antenna is also limited to 1 meter.

LOW-POWER CW TRANSMITTER

The circuit of a master-oscillator power amplifier (MOPA) CW transmitter is given in Figure 14-10. Q1 is the VFO (variable frequency oscillator) and Q2 is a Class C RF power amplifier. Meter M is a DC milliammeter which indicates Q2 collector current and which facilitates tuning of the transmitter. The telegraph key turns both transistors on and off when pressed and opened.

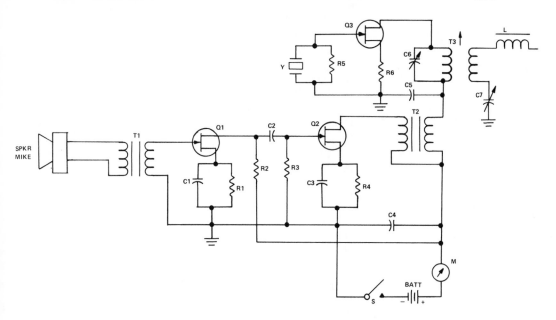

Figure 14-9. License-Free AM Transmitter

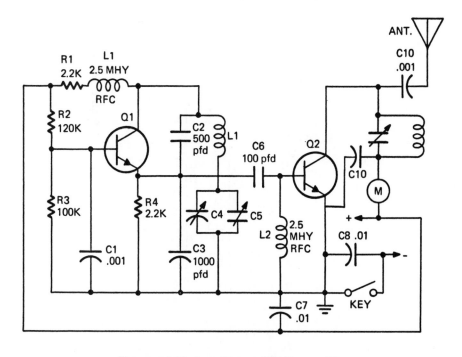

Figure 14-10. Low-Power CW Transmitter

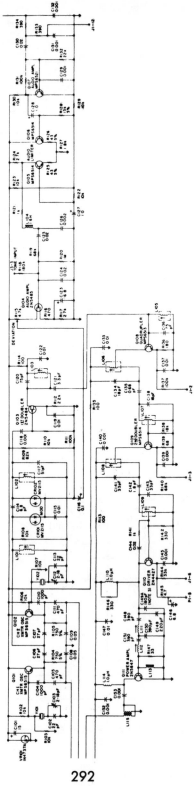

Figure 14-11. Portable FM Transceiver

PORTABLE FM TRANSCEIVER

The circuit of the transmitter of a portable FM transceiver is given in Figure 14-11. It utilizes a pair of varactor diodes (CR101-CR102) to deviate the frequency of the crystal-controlled transmitter oscillator (Q102). The microphone output is fed to the base of Q107 whose output in turn is fed through the ring-tailed pair amplitude limiter (Q104-Q106) and through a roll-off filter (L104-C126) to the gate of Q104. The deviation control (R115) is adjusted to vary the level of the audio signal applied to the varactor diodes. (*Courtesy of Hallicrafters.*)

RADIO BEACON TRANSMITTER

A single tunnel diode is used as the oscillator in the transmitter circuit given in Figure 14-12. The transmitter is modulated by an audio tone whose frequency is determined by L1 and C1. The power source is a 1.5-volt flashlight cell. This compact transmitter can be used as a radio beacon for use by boats homing in on its signal with a radio direction finder. By adding an NO SPST pushbutton in series with the positive battery terminal and common ground, it can be used as a radio control transmitter.

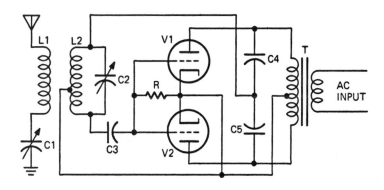

Figure 14-12. Radio Beacon Transmitter

TUNNEL DIODE FM TRANSMITTER

A tunnel diode is the active element in the FM transmitter circuit given in Figure 14-13. The transmitter carrier frequency can be varied by adjusting C2. The emitter of Q1 is coupled through C4 to the cathode of the tunnel diode. A low impedance microphone or other audio signal source is connected to J1. The power source can be a 1.34 volt mercury cell. (*Courtesy of General Electric Company.*)

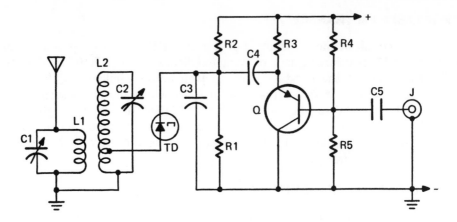

Figure 14-13. Tunnel Diode FM Transmitter

CHAPTER 15

Communications
Devices
and Systems

INTRODUCTION

Broadcasting is generally described as transmission of information, and communications as the exchange of information. Still, sometimes there are gray areas. For example, a telephone is a communications device—but when connected to a microwave network or a satellite ground station, its signal is also broadcast. And some FM broadcast stations are equipped to send private programs to different groups on an SCA (Subsidiary Carrier Authorization) channel which is not heard by persons who do not have an SCA demodulator connected to their FM receiver. This SCA channel is actually a communications service, addressed to a specific audience, such as investors, users of a background music service, etc.

Now that telegraphy is no longer in wide use, the telephone has become the primary communication medium. But the telephone would be virtually unrecognizable to Alexander Graham Bell today. Most of today's telephones are based on the rotary dial telephone, and tone-selective dialing and television receivers that also contain telephones are already reality. Since the Carterfone Decision by the FCC in 1968, users can furnish their own telephones and accessories and simply hook them up to circuits owned by the telephone company (as long as the foreign apparatus complies with FCC regulations for interface, to prevent degradation of telephone service). This single decision has accelerated the development of electronic telephones and other communication equipment.

Communications devices may be connected for *full duplex* operation or *half duplex* operation. In full duplex, both devices may talk and listen simultaneously (as in most telephone systems). In half duplex, transmission is not simultaneous—one end must listen while the other talks, and vice versa.

This chapter covers communications devices and systems that do not use space radio as the transmission medium.

AUTOMATIC TELEPHONE SET

The schematic of an automatic telephone set which also includes a dialer is given in Figure 15-1. The telephone line is connected through a modular telephone plug that interfaces the telephone line to the R (ring) and T (tip) terminals at the left side of the diagram. When the telephone handset is on hook, H-1 connects the line to bridge rectifier D1 through C1 and R1. The DC output of the rectifier, which is present when the ringing signal is received, is filtered and fed through diode D2 to the base of the PNP transistor Q1 which drives the ringer.

When the phone is off hook, H-1 connects the line through R74 to another bridge rectifier (D5) which protects the electronic circuitry from reverse polarity voltage. The level of the DC voltage is regulated by varistor D4. The DC voltage that powers the device is obtained from the common battery at the central office. When on hook, the voltage is around 50 volts but when the phone is off hook, the voltage drops to a level determined by the loop resistance of the line and the load resistance of the device. When talking into the microphone (transmitter) end of the handset, the output of TX is fed to the base of Q5 and is amplified by Q5 and Q7. The output signal developed across one winding of T1 is fed to one side of the line, and through R20, C13 and the collector-emitter path of Q2 to the other side of the line. Note that neither side of the output is fed directly to the line. Instead, the signal is fed to the line through bridge rectifier D5 which is forward-biased by the DC on the line and readily passes the low level audio signals. When

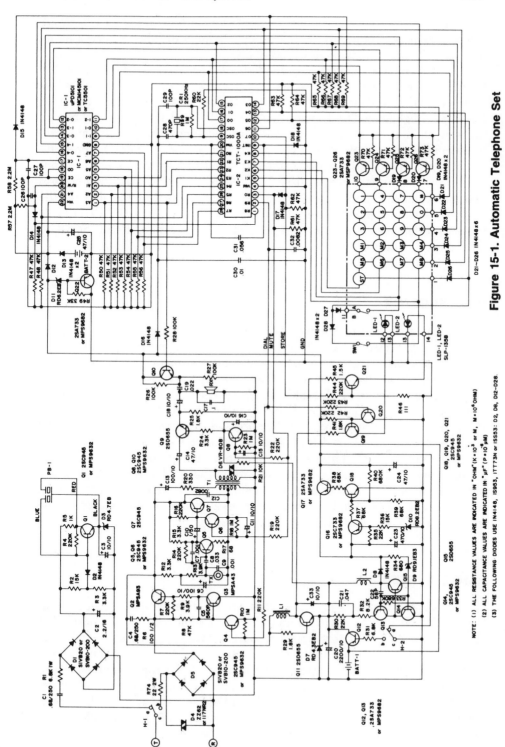

Figure 15-1. Automatic Telephone Set

listening to the receiver (earphone) end of the handset, RX is driven by Q9 whose signal is obtained from the other windings of T1. Q2 serves as an AGC circuit that is controlled by Q3 and Q4.

Any telephone number may be dialed by pressing the buttons on the keyboard. The device translates the keyboard dialing into rotary dial pulses at a speed of 10 pps or 20 pps. This enables use of the telephone equipped for rotary dialing or for Touch Tone®. You can call the same number by pressing the button, obviating the need for redialing. You can also store up to 20 numbers to enable you to call any of the stored numbers by pressing the button alongside the list of names on the directory that is on the front panel. The dialing circuitry utilizes IC1 and IC2, the keyboard and several external components. The dialing pulses are imposed on the line by the inductive kickback pulses generated by L1. A 9-volt battery (BATT-1) holds the stored numbers in the memory telephone. (*Courtesy of Radio Shack, a Division of Tandy Corp.*)

DTMF ENCODER

A dual-tone multi-frequency encoder, using the circuit given in Figure 15-2, enables transmission of Touch Tone® signals for pushbutton dialing of telephones, gaining access to and controlling automatic repeater stations, and for remote control of distant devices through a wire line or radio link.

The circuit is of a Heathkit DTMF encoder. A 12-key touch pad is connected through a quick-disconnector to the electronic circuitry as shown in the diagram. When pushbutton 1 is pressed, terminal A of the electronic circuit and terminal L1 are joined together. This causes a 697-Hz and a 1209-Hz tone to be generated and applied simultaneously to one of the inputs of IC103. The IC amplifies the tones and its output is fed through R22, the output level control, to the input of a radio transmitter or into a telephone line. Pressing any of the keys causes different two-tone combinations to be generated. (*Courtesy of the Heath Company.*)

HANDSET INTERCOM

Telephone handsets are used with the intercom amplifier depicted in Figure 15-3. It utilizes a 416 op amp. Up to 12 handsets may be paralleled across the E, G and M terminals through a 3-wire cable. Each handset is equipped with a 5000-ohm volume control for the receiver, and should have a hanger switch that automatically disconnects the microphone when in the off-hook position. Operating current for the carbon microphone is obtained from the +24-volt DC supply. Instead of a handset, a headset with a boom microphone may be used. A press-to-talk switch will prevent pickup of extraneous noise. (*Courtesy of Opamp Labs Inc.*)

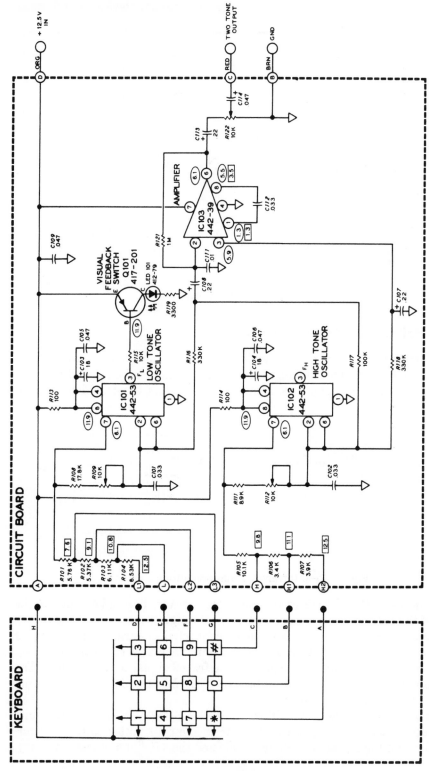

Figure 15-2. DTMF Encoder

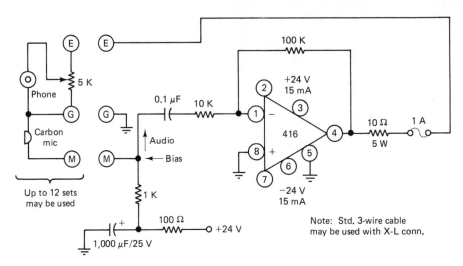

Figure 15-3. Handset Intercom

INFRARED COMMUNICATIONS SYSTEM

Infrared rays have myriad applications, including the detection of heat from vehicles and even humans. Infrared rays can also be applied to audio transmission and reception, as shown in the system illustrated by Figure 15-4.

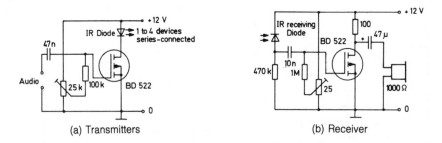

(a) Transmitters   (b) Receiver

Figure 15-4. Infrared Communications System

Figure 15-4(a) illustrates the audio transmitter. The audio signal to the gate of the FET causes a change in current through the infrared diode (which can actually be up to four such diodes connected in series). The changing current causes the IR output of the diodes to vary at a rate corresponding to the input signal, thus transmitting an infrared signal to the receiver. The 25-kohm potentiometer is used to bias the FET into the linear portion of its characteristic operating curve.

Figure 15-4(b) shows the IR receiver. The BD522 VMOS FET has quite high input resistance. Modulated light from the transmitter causes the current through the infrared receiving diode and the 470-kohm resistor to vary, and this is amplified by the FET. This circuit provides so much amplification that the output picked off through the 47 μF capacitor can be used to drive headphones, or even a loudspeaker, directly. (*Copyright © International Telephone and Telegraph Corporation.*)

LOCAL BATTERY TELEPHONE

Local battery telephones are still widely used in remote areas and by railroads for telephone service over a party line. The advantage of the local battery telephone is that no central office is required. The phones are simply bridged across the party line. As shown in Figure 15-5, a local battery telephone utilizes a dry battery (often consisting of four No. 6 dry cells). A hand-cranked magneto is used for signaling other phones on the same line or the operator when the party line is terminated at a central office. When the receiver is on hook, the phone is disconnected from the line except for the magneto and the ringer. This type of phone should not be connected to a modern subscriber loop.

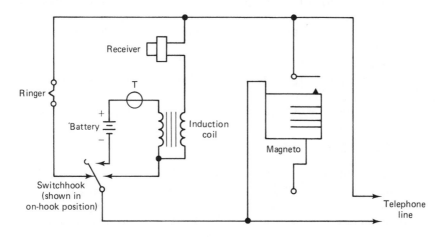

Figure 15-5. Local Battery Telephone

MODULAR TELEPHONE PLUG CONNECTIONS

The modular phone plug is rapidly replacing the 4-prong flat plug. The modular plugs are used at both ends of the handset cord and the cord that connects the phone or an accessory to the modular wall jack. The connections to the 8-position modular phonejack and plug are shown in Figure 15-6. Terminals 1 and 8 connect the phone or other telephone device to the telephone line. Terminal 1 goes to the ring side of the line and terminal 8 to the tip side. (Ring and tip refer to the three-circuit phone plug which has a tip, a ring and a sleeve.) (*Courtesy of Amphenol, an Allied Company.*)

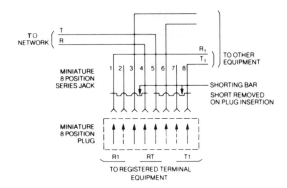

Figure 15-6. Modular Telephone Plug Connections

PULSE CODE MODULATION CODEC

In pulse code modulation, analog voice signals are digitized into a pulse coded format. This digital transmission provides cleaner speech transmission than when using conventional analog techniques, but requires more spectrum space. PCM is a very popular coding technique, especially for fiber optic transmission.

Figure 15-7 shows a diagram of a three-chip PCM codec (that is, a coder-decoder). The terminals at the left side of the diagram are connected to the "tip" and "ring" portions of the telephone line. In a typical system the coded pulse train is sampled 8,000 times per second. (*Courtesy of Motorola, Inc.*)

ROTARY DIAL TELEPHONE

The circuit of the popular 500-type telephone is given in Figure 15-8. The switches identified by "S" are parts of the hook switch assembly and those identified as "D" are the dial contacts. The varistors use the same schematic symbol as a diac. The varistors limit voltages and minimize transient peaks. When the dial is pulled, the pulsing contacts close; and when the dial is released, the contacts open and close the same number of times as the digit dialed. For example, when 8 is dialed, pulsing contacts open and close eight times. The other dial contacts remain closed while the dial is returning to its normal position to prevent the clicking of the pulsing contacts being heard.

The component identified as an induction coil is actually a hybrid transformer which matches the four-wire transmitter and receiver circuits to the two-wire telephone line. The balancing network is set to allow a little bit of sidetone so the user can hear himself or herself talk. It is set to reduce sidetone to a level that meets the operating requirements. R1 and C1 across D1 form a spark suppression circuit that minimizes clicking sounds in the receiver when dialing. RV1, RV2 and RV3 are varistors which limit the voltage and which are bidirectional.

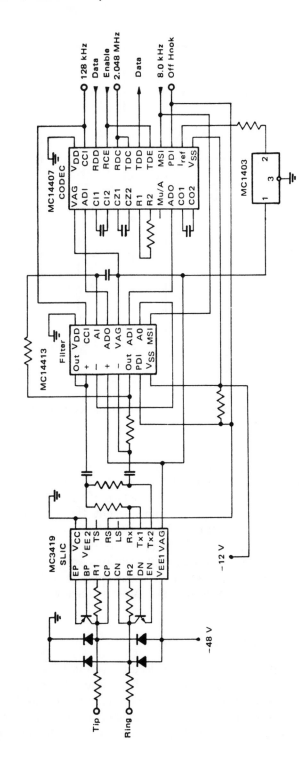

Figure 15-7. Pulse Code Modulation Codec

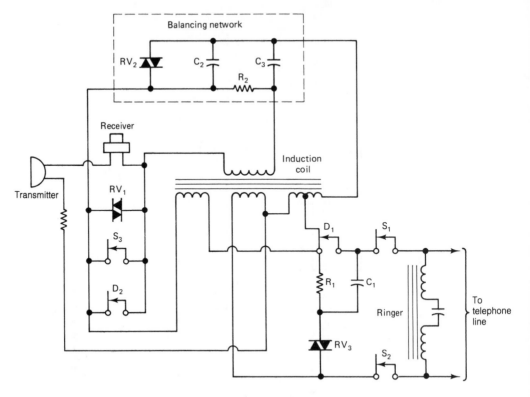

Figure 15-8. Rotary Dial Telephone

When talking into the transmitter, the DC flowing through the receiver and the bottom center winding of the induction coil varies with the speech signal. This audio signal is obtained from two of the induction coil windings. When receiving a voice signal, audio current flows through hook-switch contacts S1 and S2 and all four windings of the induction coil, and the receiver.

The Model 500 telephone is designed to work in the full duplex mode. In the half-duplex mode, two-way conversations are possible, but not simultaneously. One person talks while the other listens. In the full-duplex mode, both parties may talk simultaneously. Most phones are used in full-duplex circuits.

TELEPHONE RINGING DETECTOR

An opto-isolator is used in the circuit given in Figure 15-9 to detect the presence of a ringing signal on a telephone line. The input of the device is bridged across the telephone line with the R terminal connected to the "ring" side of the line and the T terminal connected to the "tip" side of the line. When a low-frequency AC ringing voltage is across the line, the LED in the

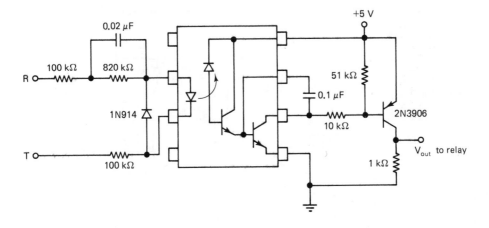

Figure 15-9. Telephone Ringing Detector

opto-isolator glows. The light is sensed by the photo detector diode whose DC output is fed to the Darlington pair (of NPN transistors) within the device. The output of the Darlington pair is fed through a 10,000-ohm resistor to the base of a PNP driver transistor whose output is connected to a relay (not shown), whose contacts may be used for actuating a bell or other alarm.

TOUCH TONE® POWER MICROPHONE

An electret (condenser) microphone and a Touch Tone® encoder are combined in the circuit given in Figure 15-10. The microphone output is amplified by the transistor and is then fed through the 5000-ohm potentiometer which is the level control. From the wiper of the level control the audio signal is fed through a 4.7-µF capacitor, a 3300-microhenry inductor, and the PTT switch to the transmitter input. An MK 5087 or an S2559 IC is used as the oscillator and frequency divider. The oscillator operates at 3.5795 MHz, the frequency determined by the crystal which is connected to pins 7 and 8 of the IC. The Touch Tone® keyboard is mounted in the microphone assembly. A built-in Zener diode is used to stabilize the operating voltage at 9.1 volts. The power source may be the 12-volt electrical system of a vehicle, a dry battery, or it may be derived from the transmitter with which the microphone is used. An LED lights when a tone is being generated. The output level of the preamplifier is −50 dB below 1 volt at 1000 Hz into a 1 megohm load. A separate 5000-ohm pot is used for setting the tone output level at the transmitter. (*Courtesy of Astatic Corp.*)

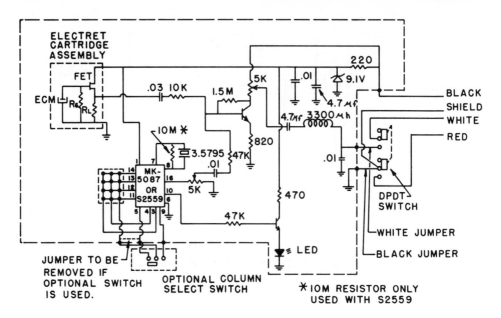

Figure 15-10. Touch Tone® Power Microphone

TOUCH TONE® TELEPHONE

The circuit of a Touch Tone (registered trademark of AT&T) telephone is given in Figure 15-11(a). Unlike the DTMF encoders described elsewhere in this book, which use RC networks or crystal-controlled oscillator and frequency dividers to generate the tones, this telephone employs L-C networks for determining the frequency of the tones. The keyboard circuit is given in Figure 15-11(b). It is connected to the taps on L1 and L2 in the phone diagram.

2-WIRE-TO-4-WIRE TERMINAL

A 2-wire-to-4-wire terminal is used to interface a two-wire duplex telephone circuit with the four-wire input and output of a radio system to enable full-duplex operation (simultaneous radio transmission and reception). Three different circuits are shown in Figure 15-12. The one shown in (a) is for use on a 600-ohm telephone circuit. It comprises two hybrid transformers and two balance networks. The one shown in (b) interfaces with a 900-ohm telephone circuit, and the one shown in (c) may be used on a 600- or 900-ohm telephone circuit. (Many people still think that a telephone line is a 600-ohm circuit. Due to the use of smaller gauge wire, most telephone lines are now rated at 900 ohms.)

The hybrid transformers, when used with the correct balance network, enable simultaneous transmission and reception over a two-wire telephone

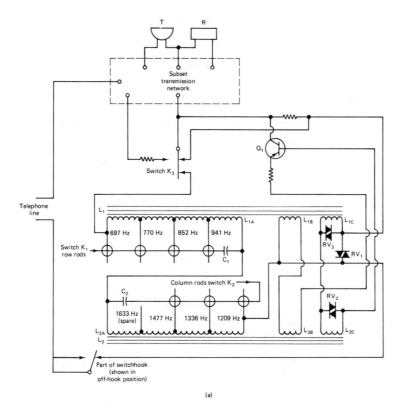

(a)

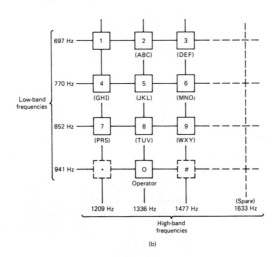

(b)

Figure 15-11. Touch Tone® Telephone

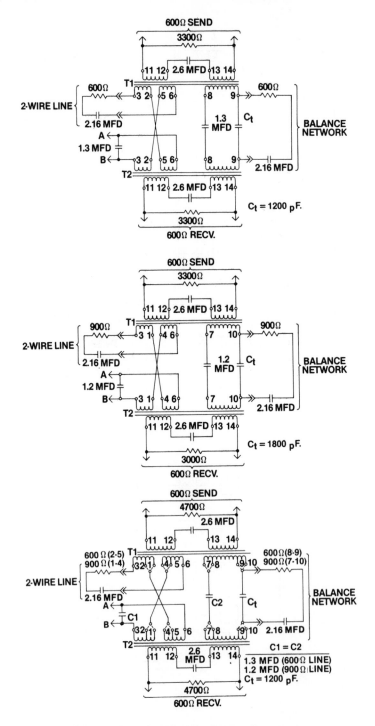

Figure 15-12. 2-Wire-To-4-Wire Terminal

line. Or a 2-wire-to-4-wire terminal may be used to convert a two-wire duplex line into a four-wire circuit. At a ham station, such a terminal can be used to interface the transmitter and receiver with the telephone line to enable full-duplex operation through a phone patch. (*Courtesy of Magnetic Controls Co.*)

WIRED INTERCOM MASTER UNIT

Here is a 5-watt output wired intercom master unit which may be operated from as high as 28 volts DC. Its speaker current will be 1.3 amperes and its quiescent supply current is only 25 milliamperes. In the circuit given in Figure 15-13 showing the use of an LM384 IC, the input is fed from an 8-ohm speaker through an 8-ohm/5000-ohm transformer (T1) which will step up the speaker output voltage 25 times. The slave speaker-microphone is also an 8-ohm type. In the listen mode, the remote speaker is connected through the section of the talk-listen switch at the left to the 8-ohm winding of T1, and the output is fed to the master unit speaker-microphone from the 8-ohm output of the amplifier. When the talk-listen switch is in the talk position, the master unit speaker-microphone is connected to the 8-ohm winding of T1 and the remote speaker is connected to the amplifier output through a 50-μF capacitor. The remote speaker should be connected to the intercom master unit through single-conductor shielded audio cable to minimize pickup of hum and RFI. The inner conductor of the cable is connected to the talk contact of the section of the talk-listen switch seen at the right, with a wire running to the listen contact of the other switch section. The cable shield is connected to pin 4 of the IC (not shown in the diagram). (*Copyright National Semiconductor Corporation.*)

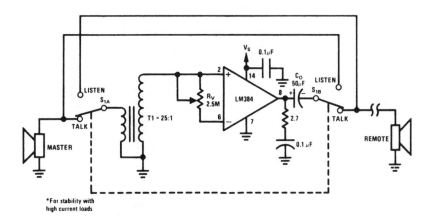

Figure 15-13. Wired Intercom Master Unit

WIRELESS INTERCOM

A wireless intercom utilizes the power line as the transmission line. It is a form of carrier current transmission system. Power line carrier systems are used by power companies for utilizing their power lines as telephone circuits. Wireless intercoms use similar techniques but use much lower power. The circuit given in Figure 15-14 is of a Realistic Selectracom wireless intercom unit. It is operable on either of two channels that can be selected with a front panel switch. The FM transmitter and receiver operate at frequencies in the 200-kHz region on which unlicensed transmitters may be operated under Part 15, FCC Rules and Regulations. The transmitter is turned on by a touch plate. Another touch plate is touched to transmit an alert tone. As can be seen in the diagram, coupling to the power line is through RF transformer T1 and coupling capacitors C1 and C2. (*Courtesy of Radio Shack, a Division of Tandy Corp.*)

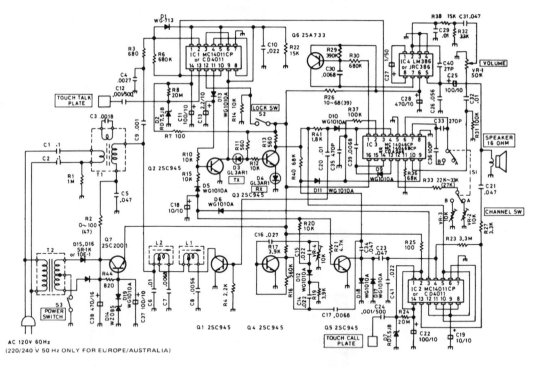

NOTE

(1) ALL RESISTANCE VALUES ARE INDICATED IN "OHM" (K=10³OHM, M=10⁶OHM)

(2) ALL CAPACITANCE VALUES ARE INDICATED IN "μF" (P=10⁻⁶μF)

Figure 15-14. Wireless Intercom

CHAPTER 16

Measuring and
Test Instruments

INTRODUCTION

From time immemorial, man has wanted to quantify the world around him—how big, how small, how far, etc.—and this is no less true in the field of electronics. By measuring the voltage and current available, the power used, or the frequency in a circuit, we can gain a better knowledge of its operation. And when, for some reason, an electronic device or circuit fails to function properly, accurate measurement is often a vital part of the test procedure.

One type of test or measurement is passive; it involves the use of test equipment that measures the circuit or device without affecting its operation. In addition, there are active types of test equipment, such as signal tracers, which alter the operation of a portion of the circuit under test.

The central problem in passive measurement is how to keep from significantly affecting the device being measured. For example, merely placing a voltmeter across a circuit seldom gives an accurate reading of voltage because the voltmeter itself draws current. The vacuum-tube voltmeter, with its high input impedance, solved this problem, but usually required a source of AC voltage to supply the high B+ needed for its tubes. A much better solution is found in today's light, portable VOMs, which use an FET input stage. The input impedance of the FET is very high—several megohms or more— so the circuit under test is affected very little by the meter.

Some passive instrumentation is permanently installed in or attached to a piece of electronic equipment to provide a readout of its operation at any time. Examples of this are a vu meter, modulation monitor, or an SWR meter. Other test equipment, both active and passive, is brought into use only when a circuit or device fails.

Active instrumentation is designed to add a signal or voltage level to a circuit. The signal which is known to be correct is injected at one point in a circuit; then the output from the circuit is monitored. Using active test instruments in this way, a troubleshooter isolates a problem to a particular stage, then narrows it down to the component level using test instruments or by simple substitution.

In order that accurate measurements be made, electronic instruments must be calibrated before being put into use, then recalibrated periodically, and this calibration should be traceable to some standard. A general rule of thumb used in industry metrology laboratories is that the tolerance of the standard used should be at least four times more accurate than the device being calibrated. Thus a voltmeter whose accuracy is ± 3 percent should be calibrated against a voltage standard that varies no more than ± 0.75 percent. Ultimately, of course, the calibration of all standards should be traceable to some ultimate beginning point, such as the standard cell at the National Bureau of Standards.

ABSORPTION WAVEMETER

A simple lumped-constant absorption wavemeter is illustrated in Figure 16-1. In a production model, different coils are plugged in to substitute for L1 in various frequency ranges. A dial associated with capacitor C1 is cali-

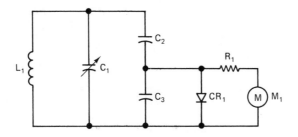

Figure 16-1. Absorption Wavemeter

brated to read frequency; the wavemeter is placed near an unknown signal and C1 is tuned until the C1-L1 tank absorbs the most RF energy—that is, it is tuned for resonance.

DIGITAL MULTIMETER

The circuit of an auto range digital multimeter is given in Figure 16-2. It utilizes discrete transistors and diodes in addition to several ICs, including a large-scale integration IC. It has a liquid crystal display, four-position function switch and a three-position on-off slide switch. When set to its DCV function, it will indicate DC volts and will automatically select the 2-, 20-, 200- or 2000-volt range. It will also select the same voltage ranges when set to measure AC volts. The input impedance on the ACV and DCV functions is 10 megohms. It will also select four different resistance ranges when set to the K ohm function. (*Courtesy of Radio Shack, a Division of Tandy Corp.*)

DIP METER

A dip meter is used to measure the frequency of a resonant circuit. The dip meter shown in Figure 16-3 uses a 2N1178 transistor as an oscillator. Coil L is one of a series of plug-in coils that give this instrument a frequency range from 3.5 to 100 MHz. Capacitor C5 is calibrated.

In operation, the instrument is turned on and held with coil L near the circuit to be measured. C5 is then tuned slowly over its range. At the frequency where the external circuit resonates and absorbs energy from the internal oscillator, the meter will "dip." Jack J allows the use of headphones so the user can hear the zero beat. (*Courtesy of RCA Solid State.*)

FREQUENCY COUNTER

The circuit of an electronic frequency counter with a range of 100 Hz to 50 MHz is given in Figure 16-4. Since this counter is a part of a transmitter tester, it utilizes a small loop to pick up the RF energy used as its input signal. It has a six-digit LED display with leading zero blanking. A quartz crystal is used as the time base. (*Courtesy of Wawasee Electronics.*)

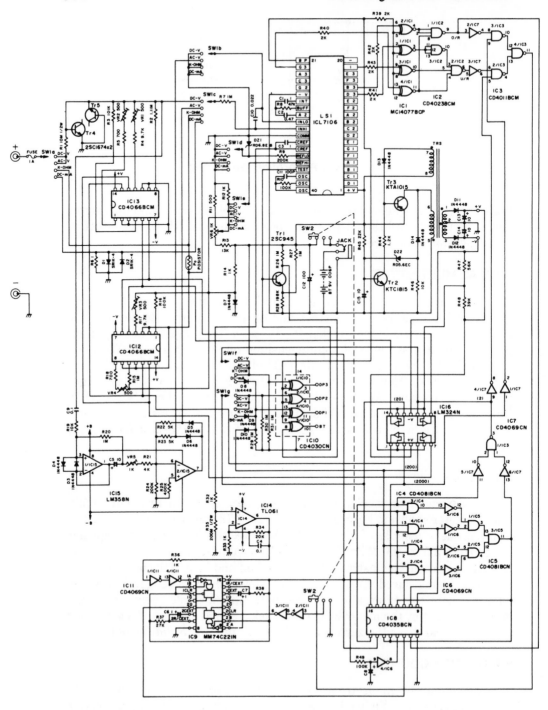

Figure 16-2. Digital Multimeter

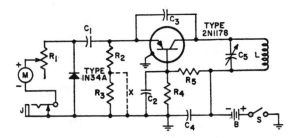

Figure 16-3. Dip Meter

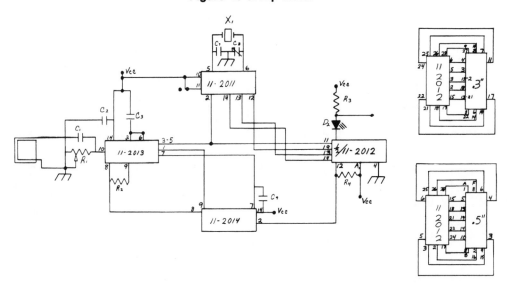

Figure 16-4. Frequency Counter

FUNCTION GENERATOR

The Hickok Model 270 function generator, whose circuit is shown in Figure 16-5, can be used for generating triangular, square and sine wave signals at frequencies within the 1-Hz to 1-MHz range at output levels adjustable from 0 to 24 volts peak-to-peak (8.5 volts RMS). An LM741CN IC (AR101) op amp is used as the oscillator, which is tuned with potentiometer R113. The output of the op amp is fed to the base of Q103 whose emitter is switch-connected to any of five potentiometers and one resistor that are used for calibrating the signal frequency; each pot covers a specific frequency range. The collector of Q103 is connected to pin 7 of Z101, a monolithic function generator IC. Pin 8 is connected to the switches that are used for selecting the frequency calibration pot.

A shunt diode (CR110) is connected between ground and the junction of R129 and CR109. R128 is connected to the range switch, a 6-volt DC source

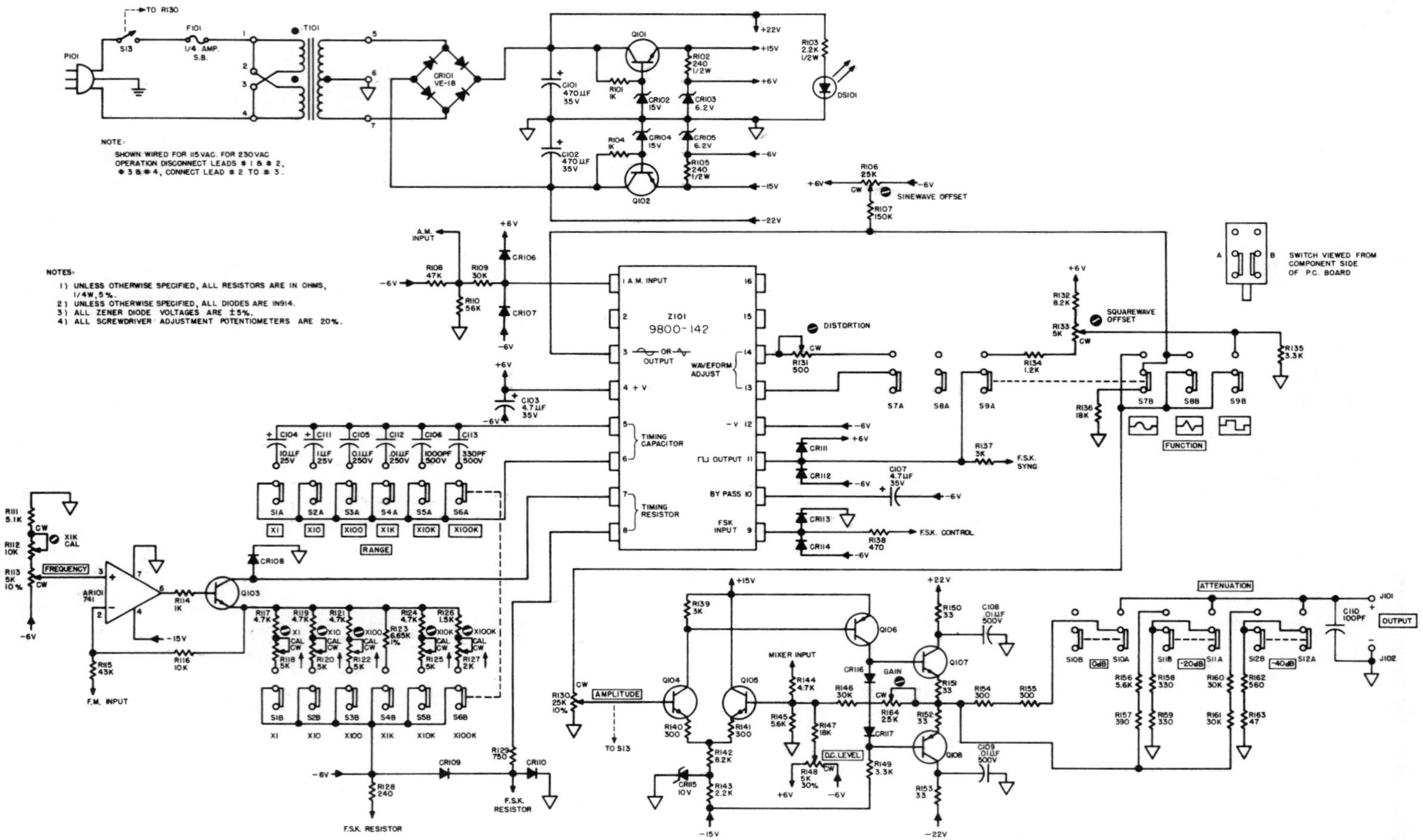

Figure 16-5. Function Generator

and the anode of CR109. An external resistor is connected to the far end of R128 and the junction of R129, CR109 and CR110 for frequency shift keying (FSK) of the output wave. The FSK input is connected to IC pin 9 through R138, which is at the junction of diodes CR113 and CR114 which limit the amplitude of the FSK signal.

The IC delivers a sine wave or a triangular wave from pin 3 to the sine wave offset pot R106 and to the function select switch. The selected signal waveform is routed through amplitude control pot R130 to one input of the differential transistor pair (Q104, Q105). The output level and signal mixing are accomplished in the circuit utilizing Q106, Q107, Q108, CR116 and CR117. The processed output signal is taken from the junction of the Q107 and Q108 emitter resistors R151 and R152. It is fed to two 500-ohm resistors (R154, R155) in series with the attenuator assembly which permits cutting insertion loss of 0 dB, 20 dB and 40 dB.

This laboratory instrument contains a built-in AC power supply which employs a full-wave bridge rectifier and two pass transistor voltage regulators, one for the +15 volt output and one for the −15 volt output. Zener diodes are used to regulate the voltage output at +6 volts and −6 volts. (*Courtesy of Hickok Electrical Instrument Company.*)

LED VU METER

LEDs are used to indicate audio level in volume units (VU) in the instrument circuit given in Figure 16-6. The circuit is powered by the 115:6.3-volt filament transformer and a full-wave rectifier whose positive DC output

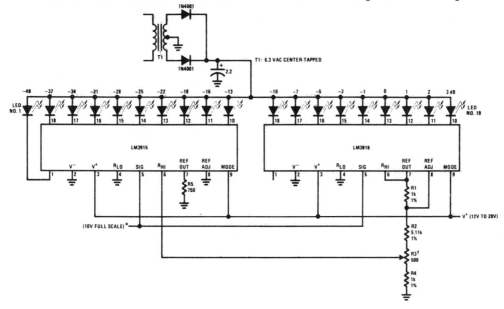

Figure 16-6. LED VU Meter

is fed to the anodes of the 19 LEDs. The audio signal to be measured is fed into pin 5 of both of the LM3915 ICs. Potentiometer R3 is adjusted to obtain a 3-dB difference between LED 11 and LED 12. The indicating range of this LED instrument is from -40 dB to $+3$ dB. (*Copyright National Semiconductor Corporation.*)

MICROPHONE TESTER

Although the quickest way to test a microphone is to substitute a new one, the output and functioning of the PTT switch of a communications microphone can be tested by using an electronic millivoltmeter and interface units as shown in Figure 16-7. The meter may be a Simpson 314 or 715 AC millivoltmeter which contains an amplifier and which has a high input impedance. The input impedance of the 314 is 10 megohms and that of the 715 is 1 megohm. The full-scale low voltage ranges of each of these meters include 0-10 millivolts, 0-30 millivolts, 0-100 millivolts, 0-300 millivolts, 0-1 volt and on up. A more expensive meter, such as the Hewlett Packard 400E/EL12, ranges from 0-1 millivolt to 0-300 volts, and its input impedance is 11 megohms.

To measure the output of a crystal, ceramic, magnetic or dynamic microphone, use the interface unit whose circuit is given in (a). Microphone jack J1 is a common type of 4-pin microphone jack and J2 is a BNC jack. The microphone plugs into J1, and J2 is connected through a shielded patch cord to the input of the meter.

To measure the output of a carbon microphone, use the interface unit whose circuit is given in (b). The microphone plugs into J1 and the meter is connected to J2. In this circuit, excitation current for the microphone is

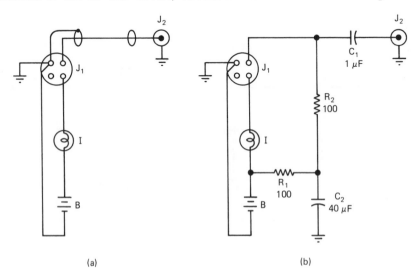

(a)   (b)

Figure 16-7. Microphone Tester

supplied by battery B through R1. The output signal is developed across R2 and is fed to the meter through C1. Capacitor C2 stabilizes the voltage at the junction of R1 and R2 and prevents R1 from acting as a "rubber band" when sound is picked up by the microphone.

In both (a) and (b), lamp I glows when the PTT switch in the microphone closes its contacts.

The connections to J1 shown are typical and do not apply to all microphones. Adaptors can be used to interface with various microphones or additional jacks of different types can be wired across J1.

MODULATION AND POWER TESTER

The circuit shown in Figure 16-8 is of a commercially available tester that checks the power output, SWR and modulation levels from an RF transmitter. An RF signal is picked off directly and applied through C4 to the wattmeter circuitry. Here diode D3 produces a DC level that is proportional to the power from the transmitter. The DC is applied through range selector switch S1 and internal calibration potentiometer VR1 to the meter, M1, which indicates power in watts.

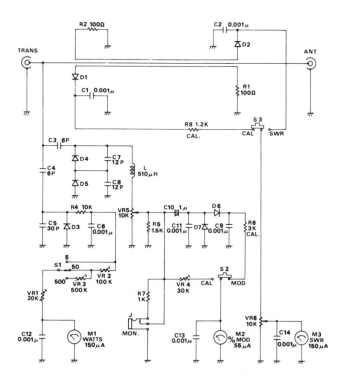

Figure 16-8. Modulation and Power Tester

An RF signal is also applied through C3 to the modulation circuit. Meter M2 is first calibrated to an unmodulated carrier in the CAL position of S2; then the modulation level is read with the switch in the MOD position. Jack J is a monitor jack that permits listening to the actual voice or other modulation.

Standing-wave ratio is measured by comparing the power at the input of a long pickup box to the power at the output. SWR is read by first calibrating the SWR meter (M3) to a carrier with S3 set to the CAL position, then switching to the SWR position. (*Courtesy of Radio Shack, a Division of Tandy Corp.*)

NOISE GENERATOR

Figure 16-9 is a schematic of a diode noise generator which can be used for aligning VHF and UHF radio receivers for maximum sensitivity. It utilizes a noise diode through which 50 microamperes of current flow when powered by a 45-volt battery. The output is fed to the input of the receiver being aligned.

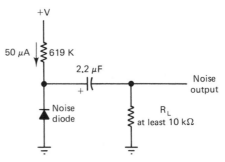

Figure 16-9. Noise Generator

RADAR DETECTOR

A radar detector is a device used in a car for sensing the presence of operating radar speed meters in the vicinity. Some people use a radar detector to enable them to unlawfully exceed the speed limit by being forewarned of a radar speed meter ahead. Others use a radar detector as a safety device.

The radar speed meter was invented by Dr. John L. Barker when he was associated with Automatic Signal in Norwalk, Connecticut. His first radar speed meter operated at a frequency around 2450 MHz. His radar speed meter utilized the "Doppler effect" for measuring the speed of a moving vehicle. When a radio signal is transmitted toward a moving target and is reflected back, the frequency of the transmitted radio signal and the reflected radio signal differ. Barker used this technology to measure speed by measuring the difference between the two frequencies and translating the result into miles per hour.

The circuit given in Figure 16-10 is of a dual-band radar detector that senses the presence of radar speed meters operating in the X band (10.5 GHz) and in the K-band (2.5 GHz). Both of these bands are in the microwave region. Although some radar detectors employ a superheterodyne receiver circuit, this one employs a tuned radio frequency (TRF) circuit. The radar signals are scooped up by a tiny horn antenna contained within the radar detector housing. The detector diodes D1 (X det) and D2 (K det) are mounted in the horn assembly. The signals intercepted by the horn antennas are detected by the applicable detector diode.

The radar detector utilizes both sections of two LM387 dual preamplifier ICs as IC1 and IC2, an LM741 op-amp as IC3, an LM567 tone decoder as IC4 and an LM78LO8A voltage regulator as IC. In addition, it utilizes both PNP and NPN transistors, 1N60 and 1N914 diodes in addition to four microwave diodes, a detector diode and an oscillator diode for each of the two bands.

The sensitivity of this radar detector, which is available in kit form, is adjusted with potentiometer R29. When the power switch is in the on position, the LED lights to indicate that the radar detector is turned on. It can be tested by adjusting R29 to trip the detector so that its indicator lamp and electronic buzzer will be actuated. Then R29 is readjusted to put it into the "armed" condition.

The radar detector may be powered by the 12-volt electrical system of the vehicle in which it is installed or from an external DC source. (*Courtesy of Radio Shack, a Division of Tandy Corp.*)

RADIO DIRECTION FINDER

The radio direction finder was invented by Dr. Fred Kolster and was improved upon by Wilbur Webb at Bendix. The Bendix radio compass is an automatic radio direction finder that was used for navigation by almost all of the world's airlines. More exotic navigation systems are now used by the airlines. However, the RDF is still the prime navigational aid used on pleasure craft. An RDF is simply a radio receiver equipped with a rotatable loop antenna which may be adjusted to provide the strongest signal or null (weakest signal) to take a bearing. An RDF may also be used for homing in on a radio signal. The circuit of a typical marine RDF is given in Figure 16-11. It is tunable through the AM and FM broadcast bands, the 1.6-4 MHz marine and amateur bands, the 150-400 kHz beacon band, and the 145-174 MHz VHF land mobile and marine bands. However, only the AM broadcast band, the 1.6-3 MHz marine band, and the long-wave beacon band are used for direction finding. This RDF has a front panel tuning meter which is used for determining the null and maximum signal positions of the loop antenna. The antenna is mounted with a compass rose calibrated in degrees of azimuth. A BFO is also provided to enable use of weak AM signals and CW signals. The RDF is powered by a self-contained battery. (*Courtesy of Pearce-Simpson.*)

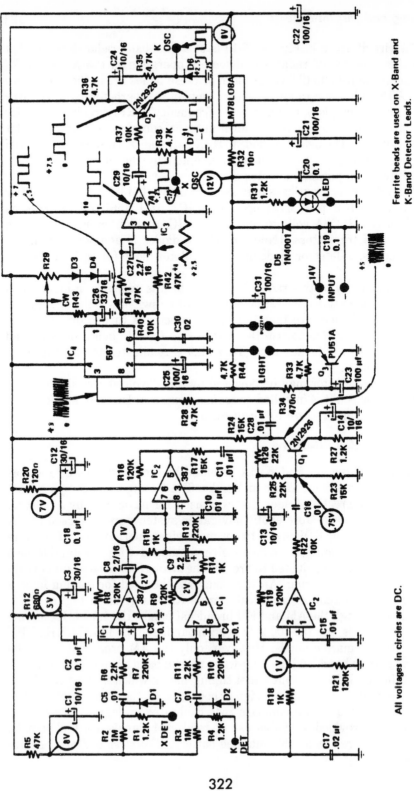

Figure 16-10. Radar Detector

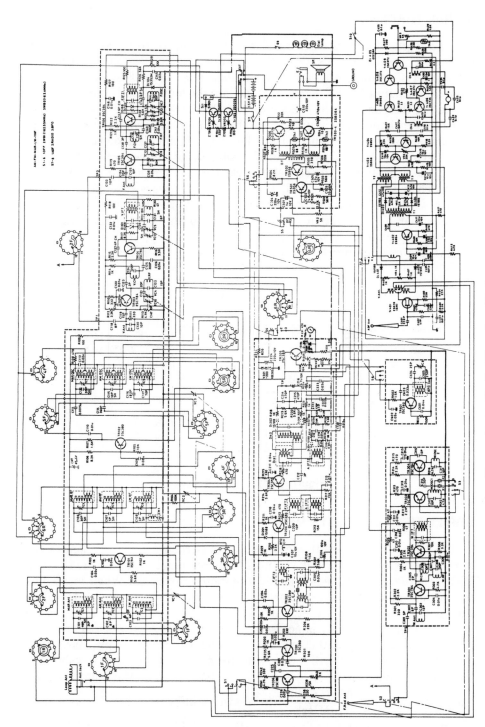

Figure 16-11. Radio Direction Finder

SIGNAL TRACER

The signal tracer whose circuit is given in Figure 16-12 can be used for tracing AF and RF signals. (For RF signal tracing, an RF detector probe is required. Its circuit is also given in the diagram.) The AF signal, direct or from the output of the RF detector, is fed into the input jack. It is capacitively coupled to the base of a BC208 transistor. The 15,000-ohm resistor in series with the transistor base circuit and the shunt capacitor form a low-pass filter. The transistor is forward-biased by the 560,000-ohm resistor connected to the collector and base. This resistor also adds negative feedback to the input stage. The output from the collector is capacitively coupled to a 47,000-ohm pot which is the volume control. The signal then passes through a voltage divider, which also functions as a low-pass filter, to the input of the TAA611b AF amplifier IC. The output of the IC is fed to an 8-ohm speaker through a 250-uF capacitor. A 9-volt transistor battery is the power source. Almost any AF power amplifier IC can be substituted for the TAA611B.

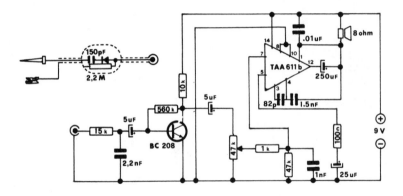

Figure 16-12. Signal Tracer

SIMPLE FIELD-STRENGTH METER

A simple field-strength meter can be constructed using two lengths of stiff copper wire, a diode, and a 50 uA DC microammeter, as shown in Figure 16-13. The diode is connected across the meter terminals. When located in the vicinity of an RF transmitter, this meter is sensitive enough to aid in tuning, antenna placement, etc.

Maximum sensitivity is achieved when the stiff copper wires are each a quarter-wave long at the operating frequency. For example, each wire should be approximately 18 inches long for the 150-MHz band, 6 inches for the 450-MHz band and 3 inches for the 800-MHz band. The length can be calculated by $0.5 \times \dfrac{5905 \times .97}{f \text{ (MHz)}}$. The answer will be in inches. For example, assume that the operating frequency (f) is 28 MHz. 5905 times .97 is 5728, and this divided by 28 MHz will give you 204.6. Then multiplying that figure

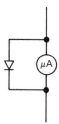

Figure 16-13. Simple Field-Strength Meter

by 0.5 will give you 102.3 inches. The figure .97 is the correction factor for the velocity factor or RF signals through wire, and it varies with the diameter of the wires. Multiplying the rest of the equation by 0.5 gives you the figure for a quarter-wave antenna element. Without multiplying by 0.5, you will have the length in inches for a half-wave antenna.

To obtain maximum sensitivity, the instrument must be held so that its antenna wires are vertical when checking the field intensity of an antenna that is vertically polarized. The same is true of horizontally polarized antennas. If the instrument is held vertically when checking the field intensity of a horizontally polarized antenna, the loss could be in excess of 20 dB.

SOUND LEVEL METER

The sound level meter whose circuit is given in Figure 16-14 can be used for measuring ambient noise levels, the sound level at various points in a hall, sound produced by a sound reinforcement system, etc. It contains a built-in dynamic microphone whose output is fed through C501 to the base of transistor TR501 (i.e., the first transistor), whose output in turn is applied across a 5-step attenuator through C503. The arm of the attenuator switch enables the user to select the 70-, 80-, 90-, 100- or 110-dB ranges. The output

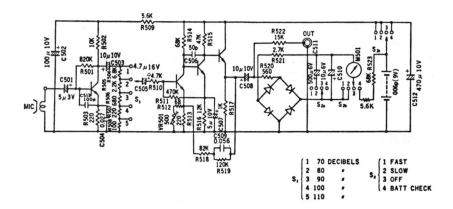

Figure 16-14. Sound Level Meter

signal at the selected tap on the attenuator switch is fed through C505 and R510 to the base of TR502. Output from this transistor's collector is direct-coupled to the base of TR503, and through C506 to the base of TR504, an emitter follower whose output is fed through C508 and R520 to the full-wave bridge rectifier which rectifies the amplified audio and noise signals. The negative DC output of the bridge rectifier is fed through R521 to one side of meter M501. The positive output terminal of the bridge rectifier is connected directly to the meter terminal. Switch S2 is a 3P4T rotary switch that can be set to any of the following four positions: (1) fast response, (2) slow response, (3) power off and (4) battery check. In the slow position of S2, a 500-μF capacitor (C511) is shunted across the meter to slow down its response time. In the battery check position, the meter is connected across the integral 9-volt transistor battery through a 5600-ohm resistor and a 68,000-ohm resistor. A headset can be plugged into the OUT jack to enable hearing sounds picked up by the microphone. (*Courtesy of Radio Shack, a Division of Tandy Corp.*)

SWR MONITORING CIRCUIT

When a transmitter is equipped with a directional coupler that enables measurement of forward power and reflected power, two signals are available for measurement. Most meters connected to the two voltage sources are calibrated in SWR (standing wave ratio). When the antenna and its trans-mission line impedances are exactly matched, SWR is 1:1. In actual practice, the SWR is actually closer to 1.5:1. With an SWR of 1.5:1, 4 percent of the available power is reflected back from the antenna to the transmitter. When SWR exceeds 2.5:1, the final RF stage transistor or tube might be damaged. Therefore, some transmitters are equipped with an SWR alarm circuit which sounds an alarm, lights a lamp or shuts off the power to the final RF stage when SWR exceeds its safe value. The circuit given in Figure 16-15 is an example of an automatic SWR circuit. The forward power signal from the directional coupler is fed to the FOR input terminal and the reflected power signal is fed to the REF terminal. The DC operating voltage is fed to the B terminal and a meter is connected to the M terminal. An alarm or lamp is driven from the ALERT terminal. Each input signal is amplified and then the ratio is determined. When the reflected power is greater than, say, 10 percent of the forward power, the SWR monitoring device will sound an alarm. The threshold can be adjusted with the potentiometers. (*Courtesy of Quintron Corporation.*)

TONE SQUELCH TESTER

The circuit of a test accessory, the Toner, is given in Figure 16-16(a). It is intended to be used for testing tone squelch systems used in land mobile radio applications. The Toner is an accessory for use with frequency counters, which facilitates the measurement of frequencies in the range from 50 Hz

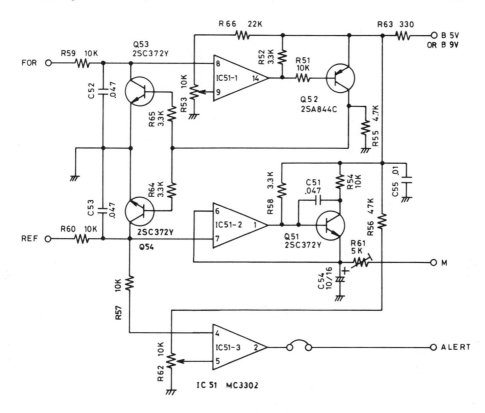

Figure 16-15. SWR Monitoring Circuit

to 5,000 Hz. Ordinary frequency counters require a ten-second count to measure frequencies to a resolution of 0.1 Hz, and a one-second count to measure frequencies to a resolution of 1Hz. Circuitry in the Toner multiplies incoming frequencies by a factor of 10 or 100, permitting faster high-resolution measurements. The Toner will provide a multiplication factor of ten for frequencies in the range of 50 Hz to 5,000 Hz. In the range of 50 to 300 Hz, the Toner will provide a multiplication factor of 100. When switched to the range for multiplication of 100, the Toner provides a low-pass filter, with very high attenuation above 300 Hz. This filter permits the measurement of tone squelch frequencies in the presence of voice and noise.

A block diagram of the Toner is given in Figure 16-16(b). The input circuits are connected to the source of the tone frequency to be measured, and the output (labeled "Counter") is connected to a frequency counter. When the function switch labeled "Direct" is pushed in, the input terminal is connected to the counter terminal, and the frequency counter receives the input signal directly, permitting the user to leave the Toner connected to the counter for measurements in which its functions are not required.

When the function switch labeled "× 10" is pushed in, the incoming tone signal is amplified by the input amplifier, then further amplified and

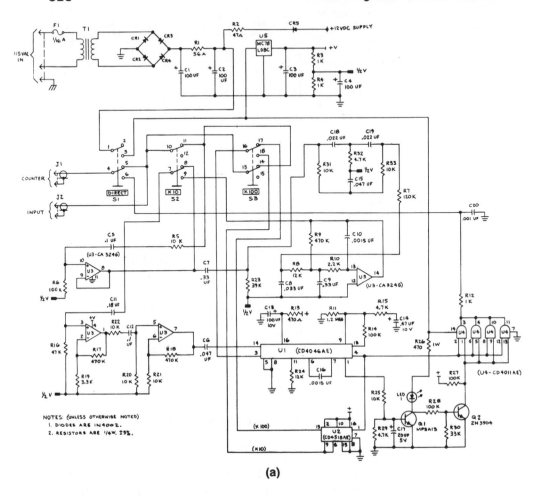

(a)

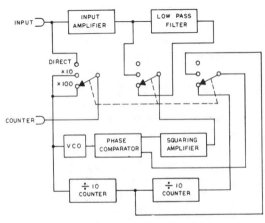

(b)

Figure 16-16. Tone Squelch Tester

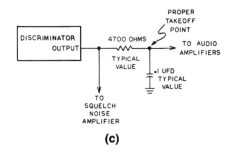

Figure 16-16. Tone Squelch Tester (Continued)

limited by the squaring amplifier. The output of the squaring amplifier is a square wave, having the same frequency as the input tone signal, and this square wave is fed into one of the inputs of the phase comparator. The phase comparator controls the frequency of the voltage-controlled oscillator. The output of the VCO is fed to a divide-by-10-counter circuit, and the output of this counter is fed back into the other input of the phase comparator. The action of the phase comparator and the VCO is such that the output of the VCO becomes locked to a frequency exactly 10 times the frequency of the input tone signal. This VCO output is then delivered to the frequency counter input. The frequency counter will show a frequency count exactly 10 times the frequency of the incoming tone signal.

The Toner can be used for measuring tone frequencies off-the-air, using a monitor receiver. Higher frequency tones can be satisfactorily measured by simply connecting a test lead between the input lead and the monitor receiver's loudspeaker. Tone squelch frequencies are greatly attenuated before they reach the loudspeakers, and it is better to make a special connection inside the receiver if off-the-air measurement of tone squelch frequencies is desired. The proper take-off point for the connection is just after the audio de-emphasis network which usually follows the FM detector (discriminator). Since many monitors have similar circuits, the diagram (Fig. 16-16(c)) will help locate this point. The Toner utilizes five ICs and two NPN transistors. The device is designed for operation from a standard 60-Hz 117-volt AC source. (*Courtesy of Helper Instrument Company.*)

Semiconductor Symbols and Abbreviations

The letter symbols and abbreviations in this glossary are found in general use in specification sheets and catalogs, and closely match those recommended by the National Electrical Manufacturers Association (NEMA) and the Joint Electron Device Engineering Council (JEDEC) of the Electronic Industries Association (EIA).

A, a	Anode
B, b	Base
C, c	Collector
C_{cb}	Interterminal capacitance, collector to base
C_{ce}	Interterminal capacitance, collector to emitter
C_{ds}	Capacitance, drain to source
C_{du}	Capacitance, drain to substrate
C_{eb}	Capacitance, emitter to base
C_{ibo}	Common-base open-circuit input capacitance
C_{ibs}	Common-base short-circuit input capacitance
C_{ieo}	Common-emitter open circuit input capacitance
C_{ies}	Common-emitter short circuit input capacitance
C_{iss}	Common-source short-circuit input capacitance
C_{obo}	Common-base open-circuit output capacitance
C_{obs}	Common-base short-circuit output capacitance
C_{oeo}	Common-emitter open-circuit output capacitance
C_{oes}	Common-emitter short-circuit output capacitance
C_{oss}	Common-source short-circuit output capacitance
C_{rbs}	Common-base short-circuit reverse transfer capacitance
C_{rcs}	Common-collector short-circuit reverse transfer capacitance
C_{res}	Common-emitter short-circuit reverse transfer capacitance
C_{rss}	Common-source short-circuit reverse transfer capacitance
C_{tc}	Collector depletion-layer capacitance
C_{te}	Emitter depletion-layer capacitance
D, d	Drain
E, e	Emitter
η	Intrinsic standoff ratio
f_{hfb}	Common-base small-signal short-circuit forward current transfer ratio cutoff frequency
f_{hfc}	Common-collector small-signal short-circuit forward current transfer ratio cutoff frequency
f_{max}	Maximum frequency of oscillation
F_T	Transition frequency (frequency at which common-emitter small-signal forward current transfer ratio would reach unity)
G, g	Gate
g_{fs}	Common-source small-signal forward transfer conductance
g_{is}	Common-source small-signal input conductance
g_{MB}	Common-base static transconductance
g_{MC}	Common-collector static transconductance
g_{ME}	Common-emitter static transconductance
g_{os}	Common-source small-signal output conductance
G_{PB}	Common-base large-signal insertion power gain

G_{pb}	Common-base small-signal insertion power gain
G_{PC}	Common-collector large-signal insertion power gain
G_{pc}	Common-collector small-signal insertion power gain
G_{PE}	Common-emitter large-signal insertion power gain
G_{pe}	Common-emitter small-signal insertion power gain
C_{pg}	Common-gate small-signal insertion power gain
C_{ps}	Common-source small-signal insertion power gain
g_{rs}	Common-source small-signal reverse transfer conductance
G_{TB}	Common-base large-signal transducer power gain
G_{tb}	Common-base small-signal transducer power gain
G_{TC}	Common-collector large-signal transducer power gain
G_{tc}	Common-collector small-signal transducer power gain
G_{TE}	Common-emitter large-signal transducer power gain
G_{te}	Common-emitter small-signal transducer power gain
G_{tg}	Common-gate small-signal transducer power gain
G_{ts}	Common-source small-signal transducer power gain
h_{FB}	Common-base static forward current transfer ratio
h_{fb}	Common-base small-signal short-circuit forward current transfer ratio
h_{FC}	Common-collector static forward current transfer ratio
h_{fc}	Common-collector small-signal short-circuit forward current transfer ratio
h_{FE}	Common-emitter static forward current transfer ratio
h_{fe}	Common-emitter small-signal short-circuit forward current transfer ratio
h_{FEL}	Inherent large-signal forward current transfer ratio
h_{IB}	Common-base static input resistance
h_{ib}	Common-base small-signal short-circuit input impedance
h_{IC}	Common-collector static input resistance
h_{ic}	Common-collector small-signal short-circuit input impedance
h_{IE}	Common-emitter static input resistance
h_{ie}	Common-emitter small-signal short-circuit input impedance
h_{ob}	Common-base small-signal open-circuit output admittance
h_{oc}	Common-collector small-signal open-circuit output admittance
h_{oe}	Common-emitter small-signal open-circuit output admittance
h_{rb}	Common-base small-signal open-circuit reverse voltage transfer ratio
h_{rc}	Common-collector small-signal open-circuit reverse voltage transfer ratio
h_{re}	Common-emitter small-signal open-circuit reverse voltage transfer ratio

I_B	Dc current through base terminal
I_b	Ac component (rms) of collector-terminal current
i_B	Instantaneous total value of base-terminal current
I_{BEV}	Base cutoff current, dc
$I_{B2(mod)}$	Interbase modulated current
I_C	Dc current through collector terminal
I_c	Ac component (rms) of collector terminal current
i_C	Instantaneous total value of collector-terminal current
I_{CBO}	Collector cutoff current (dc), emitter open
I_{CEO}	Collector cutoff current (dc), base open
I_{CER}	Collector cutoff current (dc), specified resistance between base and emitter
I_{CES}	Collector cutoff current (dc), base shorted to emitter
I_{CEV}	Collector cutoff current (dc), specified voltage between base and emitter
I_{CEX}	Collector cutoff current (dc), specified circuit between base and emitter
I_D	Dc drain current
$I_{D(off)}$	Drain cutoff current
$I_{D(on)}$	Drain current, device on
I_{DSS}	Zero-gate-voltage drain current
I_E	Dc current at emitter terminal
I_e	Ac component (rms) of emitter-terminal current
i_E	Instantaneous total value of emitter-terminal current
I_{EBO}	Emitter cutoff current (dc), collector open
I_{EB20}	Emitter reverse current
$I_{EC(ofs)}$	Emitter-collector offset current
I_{ECS}	Emitter cutoff current (dc), base short-circuited to collector
$I_{E1E2(off)}$	Emitter cutoff current
I_F	Dc forward current for diode (For signal and rectifier diodes, no alternating component)
I_f	Alternating component (rms values) of forward current
i_F	Instantaneous total forward current
$I_{F(AV)}$	Dc forward current (with alternating component)
I_{FM}	Maximum (peak) total forward current
$I_{F(OV)}$	Overload forward current
I_{FRM}	Maximum (peak) repetitive forward current
$I_{F(RMS)}$	Total rms forward current
I_{FSM}	Maximum (peak) surge forward current
I_G	Dc gate current
I_{GF}	Forward gate current
I_{GR}	Reverse gate current
I_{GSS}	Reverse gate current, drain short-circuited to source
I_{GSSF}	Forward gate current, drain short-circuited to source
I_{GSSR}	Reverse gate current, drain short-circuited to source
I_f	Inflection-point current

I_O	Average forward current, 180° conduction angle, 60-Hz half sine wave
I_p	Peak-point current
I_R	Dc reverse current for diode (For signal and rectifier diodes, no alternating component)
I_r	Ac component (rms) of reverse current
i_R	Instantaneous total reverse current
$I_{R(AV)}$	Dc reverse current (with alternating component)
I_{RM}	Maximum (peak) total reverse current
I_{RRM}	Maximum (peak) repetitive reverse current
$I_{R(RMS)}$	Total rms reverse current
I_{RSM}	Maximum (peak) surge reverse current
I_S	Dc source current
I_{SDS}	Zero-gate-voltage source current
$I_{S(off)}$	Source cutoff current
I_V	Valley-point current
I_Z	Regulator current, dc reference current
I_{ZK}	Regulator current, dc reference current near breakdown knee
I_{ZM}	Regulator current, dc maximum rated reference current
K, k	Cathode
L_c	Conversion loss
M	Figure of merit
NF_o	Overall noise figure
NR_o	Output noise ratio
P_{BE}	Dc power input to base, common emitter
p_{BE}	Instantaneous total power input to base, common emitter
P_{CB}	Dc power input to collector, common base
p_{CB}	Instantaneous total power input to collector, common base
P_{CE}	Dc power input to collector, common emitter
p_{CE}	Instantaneous total power input to collector, common emitter
P_{EB}	Dc power input to emitter, common base
p_{EB}	Instantaneous total power input to emitter, common base
P_F	Forward dc power dissipation (no alternating component)
p_F	Instantaneous total forward power dissipation
$P_{F(AV)}$	Forward dc power dissipation (with alternating component)
P_{FM}	Maximum (peak) total forward power dissipation
P_{IB}	Common-base large-signal input power
P_{ib}	Common-base small-signal input power
P_{IC}	Common-collector large-signal input power
P_{ic}	Common-collector small-signal input power
P_{IE}	Common-emitter large-signal input power
P_{ie}	Common-emitter small-signal input power
P_{OB}	Common-base large-signal output power
P_{ob}	Common-base small-signal output power

P_{OC}	Common-collector large-signal output power
P_{oc}	Common-collector small-signal output power
P_{OE}	Common-emitter large-signal output power
P_{oe}	Common-emitter small-signal output power
P_R	Dc reverse power dissipation (no alternating component)
p_R	Instantaneous total reverse power dissipation
$P_{R(AV)}$	Dc reverse power dissipation (with alternating component)
P_{RM}	Maximum (peak) total reverse power dissipation
P_T	Total nonreactive power input to all terminals
p_T	Instantaneous total nonreactive power input to all terminals
Q_S	Stored charge
r_{BB}	Interbase resistance
$r_b C_c$	Time constant, collector-base
$r_{CE(sat)}$	Collector-to-emitter saturation resistance
$r_{DS(on)}$	Static drain-source on-state resistance
$r_{ds(on)}$	Small-signal drain-source on-state resistance
$Re(h_{ie})$	Real part of common-emitter small-signal short-circuit input impedance
$Re(h_{oe})$	Real part of common-emitter small-signal short-circuit input impedance
$r_{e1e2(on)}$	Small-signal emitter-emitter on-state resistance
r_i	Dynamic resistance at inflection point
R_θ	Thermal resistance
$R_{\theta CA}$	Case to ambient thermal resistance
$R_{\theta JA}$	Junction to ambient thermal resistance
$R_{\theta JC}$	Junction to case thermal resistance
S, s	Source
T_A	Ambient or free-air temperature
T_C	Case temperature
t_d	Delay time
$t_{d(off)}$	Turn-off delay time
$t_{d(on)}$	Turn-on delay time
t_f	Fall time
t_{fr}	Forward recovery time
T_j	Junction temperature
t_{off}	Turn-off time
t_{on}	Turn-on time
t_p	Pulse time
t_r	Rise time
t_{rr}	Reverse recovery time
t_s	Storage time
TSS	Tangenital signal sensitivity
T_{stg}	Storage temperature
t_w	Pulse average time
U, u	Bulk (substrate)

V_{BB}	Dc base supply voltage
V_{BC}	Average or dc voltage, base to collector
V_{bc}	Instantaneous value of alternating component of base-emitter voltage
V_{BE}	Average or dc voltage, base to emitter
v_{be}	Instantaneous value of alternating component of base-emitter voltage
$V_{(BR)}$	Dc breakdown voltage
$v_{(BR)}$	Breakdown voltage (instantaneous total)
$V_{(BR)CBO}$	Collector-base breakdown voltage, emitter open
$V_{(BR)CEO}$	Collector-emitter breakdown voltage, base open
$V_{(BR)CER}$	Collector-emitter breakdown voltage, resistance between base and emitter
$V_{(BR)CES}$	Collector-emitter breakdown voltage, base shorted to emitter
$V_{(BR)CEV}$	Collector-emitter breakdown voltage, specified voltage between base and emitter
$V_{(BR)CEX}$	Collector-emitter breakdown voltage, specified circuit between base and emitter
$V_{(BR)EBO}$	Emitter-base breakdown voltage, collector open
$V_{(BR)ECO}$	Emitter-collector breakdown voltage, base open
$V_{(BR)E1E2}$	Emitter-emitter breakdown voltage
$V_{(BR)GSS}$	Gate-source breakdown voltage
$V_{(BR)GSSF}$	Gate-source forward breakdown voltage
$V_{(BR)GSSR}$	Gate-source reverse breakdown voltage
V_{B2B1}	Interbase voltage
V_{CB}	Collector-to-base average or dc voltage
v_{cb}	Instantaneous value of alternating component of collector-to-base voltage
$V_{CB(fl)}$	Dc open-circuit collector-to-base voltage (floating potential)
V_{CBO}	Collector-to-base voltage, emitter open
V_{CC}	Dc collector supply voltage
V_{CE}	Collector-to-emitter average or dc voltage
V_{ce}	Instantaneous value of alternating component of collector-to-emitter voltage
$V_{CE(fl)}$	Dc open-circuit collector-to-emitter voltage (floating potential)
V_{CEO}	Dc collector-to-emitter voltage, base open
$V_{CE(ofs)}$	Collector-to-emitter offset voltage
V_{CER}	Dc collector-to-emitter voltage, resistance between base and emitter
V_{CES}	Dc voltage between collector and emitter, base shorted to emitter
$V_{CE(sat)}$	Dc saturation voltage, collector to emitter

V_{CEV}	Dc collector-to-emitter voltage, specified voltage between base and emitter
V_{CEX}	Dc collector-to-emitter voltage, specified circuit between base and emitter
V_{DD}	Dc drain supply voltage
V_{DG}	Drain-to-gate voltage
V_{DS}	Drain-to-source voltage
$V_{DS(on)}$	Drain-to-source on-state voltage
V_{DU}	Drain-to-substrate voltage
V_{EB}	Emitter-to-base average or dc voltage
V_{eb}	Instantaneous value of alternating component of emitter-to-base voltage
$V_{EB(fl)}$	Dc open-circuit emitter-to-base voltage (floating potential)
V_{EBO}	Dc emitter-to-base voltage, collector open
$V_{EB(sat)}$	Emitter saturation voltage
V_{EC}	Emitter-to-collector average or dc voltage
v_{ec}	Instantaneous value of alternating component of emitter-to-collector voltage
$V_{EC(fl)}$	Dc open-circuit emitter-to-collector voltage (floating potential)
$V_{EC(ofs)}$	Emitter-to-collector offset voltage
V_{EE}	Dc emitter supply voltage
V_F	Dc forward voltage for voltage-regulator and voltage reference diodes. Dc forward voltage with no alternating component for signal and rectifier diodes.
V_f	Alternating component (rms value) of forward voltage
V_F	Instantaneous total forward voltage
$V_{F(AV)}$	Dc forward voltage (with alternating component)
V_{FM}	Maximum (peak) total forward voltage
$V_{F(RMS)}$	Total rms forward voltage
V_{GG}	DC gate supply voltage
V_{GS}	Gate-to-source voltage
V_{GSF}	Gate-to-source forward voltage
$V_{GS(off)}$	Gate-to-source cutoff voltage
V_{GSR}	Gate-to-source reverse voltage
$V_{GS(th)}$	Gate-to-source threshold voltage
V_{GU}	Gate-to-substrate voltage
V_I	Inflection-point voltage
V_{OB1}	Base-1 peak voltage
V_P	Peak-point voltage
V_{PP}	Projected peak-point voltage
V_R	Dc reverse voltage for voltage-regulator and voltage-reference diodes. Dc reverse voltage with no alternating component for signal and rectifier diodes.
V_r	Alternating component (rms value) of reverse voltage

v_R	Instantaneous total reverse voltage
$V_{R(AV)}$	Dc reverse voltage (with alternating component)
V_{RM}	Maximum (peak) total reverse voltage
V_{RRM}	Repetitive peak reverse voltage
$V_{R(RMS)}$	Total rms reverse voltage
V_{RSM}	Nonrepetitive peak reverse voltage
V_{RT}	Reach-through voltage
V_{RWM}	Working peak reverse voltage
V_{SS}	Dc source supply voltage
V_{SU}	Source-to-substrate voltage
V_{TO}	Threshold voltage
V_V	Valley point voltage
V_Z	Regulator voltage, dc reference voltage
V_{ZM}	Regulator voltage, dc reference voltage (at maximum rated current)
y_{fb}	Common-base small-signal short-circuit forward transfer admittance
y_{fc}	Common-collector small-signal short-circuit forward transfer admittance
y_{fe}	Common-emitter small-signal short-circuit forward transfer admittance
y_{fs}	Common-source small-signal short-circuit forward transfer admittance
$y_{fs(imag)}$	Common-source small-signal forward transfer susceptance
$y_{fs(real)}$	Common-source small-signal forward transfer conductance
y_{ib}	Common-base small-signal short-circuit input admittance
y_{ic}	Common-collector small-signal short-circuit input admittance
y_{ie}	Common-emitter small-signal short-circuit input admittance
$y_{ie(imag)}$	Imaginary part of common-emitter small-signal short-circuit input admittance
$y_{ie(real)}$	Real part of common-emitter small-signal short-circuit input admittance
y_{is}	Common-source small-signal short-circuit input admittance
$y_{is(imag)}$	Common-source small-signal input susceptance
$y_{is(real)}$	Common-source small-signal input conductance
y_{ob}	Common-base small-signal short-circuit output admittance
y_{oc}	Common-collector small-signal short-circuit output admittance
y_{oe}	Common-emitter small-signal short-circuit output admittance
$y_{oe(imag)}$	Imaginary part of common-emitter small-signal short-circuit output admittance

$y_{oe(real)}$	Real part of common-emitter small-signal short-circuit output admittance
y_{os}	Common-source small-signal short-circuit output admittance
$y_{os(imag)}$	Common-source small-signal short-circuit output susceptance
$y_{os(real)}$	Common-source small-signal short-circuit output conductance
y_{rb}	Common-base small-signal short-circuit reverse transfer admittance
y_{rc}	Common-collector small-signal short-circuit reverse transfer admittance
y_{re}	Common-emitter small-signal short-circuit reverse transfer admittance
y_{rs}	Common-source small-signal short-circuit reverse transfer admittance
$y_{rs(imag)}$	Common-source small-signal reverse transfer susceptance
$y_{rs(real)}$	Common-source small-signal reverse transfer conductance
z_{if}	Intermediate-frequency impedance
z_m	Modulator-frequency load impedance
z_{rf}	Radio-frequency impedance
$Z_{\theta JA(t)}$	Junction-to-ambient transient thermal impedance
$Z_{\theta JC(t)}$	Junction-to-case transient thermal impedance
$Z_{\theta(t)}$	Transient thermal impedance
z_v	Video impedance
z_z	Regulator impedance, reference impedance (small-signal at I_Z)
z_{zk}	Regulator impedance, reference impedance (small-signal at I_{ZK})
z_{zm}	Regulator impedance, reference impedance (small-signal at I_{ZM})

INDEX

(Boldface page numbers refer to the actual headings of the circuits in this book.)

341